JAPAN'S SEARCH FOR OIL

A Case Study on Economic Nationalism and International Security

Yuan-li Wu

HOOVER INSTITUTION PRESS

Stanford University, Stanford, California

Hoover Institution Publication 165

The potentially most serious international consequences could occur in relations between North America, Europe, and Japan. If the energy crisis is permitted to continue unchecked, some countries will be tempted to secure unilateral benefit through separate arrangements with producers at the expense of the collaboration that offers the only hope for survival over the long term. . . . Traditional patterns of policy may be abandoned because of dependence on a strategic commodity. Even the hopeful process of easing tensions with our adversaries would suffer because it has always presupposed the political unity of the Atlantic nations and Japan.

HENRY A. KISSINGER
"Energy Crisis: Strategy for Cooperative Action"
Chicago—November 14, 1974

Adjustments in U.S. economic policies and a new policy toward China in 1971 led to painful but transitory misunderstandings to which— let us be frank—our own tactics contributed. We have learned from experience; these strains are behind us; our policies are moving in harmony in these areas; our consultations on all major issues are now close, frequent, and frank.

HENRY A. KISSINGER
speech before the Japan Society
New York—June 18, 1975

CONTENTS

LIST OF TABLES

PREFACE

The present study is another installment in the author's research on economic nationalism and international security. A previous study—*The Strategic Land Ridge*, published by the Hoover Institution in 1975—deals with nationalism in its political manifestation and the microprocess of perception and response among countries in their respective policymaking, with special reference to the China policy of four Southeast Asian countries. Here the focus is on Japan and the uncertain effects of economic nationalism. The analysis deals with a situation where an intelligent solution based on international—especially U.S. and Japanese—cooperation may lag because of misunderstanding as well as lack of understanding, lethargy, and other national preoccupations. It is addressed in particular to those Americans and Japanese who wish to see the postwar alliance of the two countries endure. It attempts to cast light on some potential trouble spots so that they can be identified, if not removed, in advance. To repeat, the purpose of the study is not to forecast, but to point to contingencies in order better to avoid them even though no irremediable contingency in U.S.-Japanese relations has arisen so far.

The author is grateful to the Earhart Foundation and the Hoover Institution for their unfailing, invaluable support. Thanks are due especially to many friends in Japan, including William Royce of SRI-Tokyo, James Stewart of the Asia Foundation, Tokyo Office, John Hart at the American Chamber of Commerce (Tokyo), Don Westmore at the U.S. Embassy (Japan), and many scholars in the Institute of Energy Economics and MITI (Ministry of International Trade and Industry), as well as corporate officials in the Japanese oil industry who have been most generous in sharing with him their information and views but whose names cannot all be mentioned here. He wishes to thank H. C. Ling, Chi-yi Hsiang, and Akira Yasuoka for their indispensable assistance in gathering data, and Cathie Yeardye for painstakingly typing and correcting several versions of this manuscript. In revising the manuscript the author has benefitted greatly from the comments of a number of persons including, in particular, John K. Emmerson at Stanford University and Joseph Yaeger of the Brookings Institution who both read an earlier draft, Chan Yang Bang at the

University of San Francisco, and members of a Federal Reserve Bank of San Francisco seminar where a preliminary paper on the subject was first aired. All errors of fact and judgment are, of course, strictly the author's responsibility.

YUAN-LI WU

Menlo Park, California
1977

INTRODUCTION

This study focuses on Japan's experience in the 1973–74 oil crisis which was triggered by an Arab threat of total boycott and a sudden phenomenal increase in the price of oil. In particular, our attention will be centered on the nature and manner of Japan's response to this economic threat and on some of the potentially serious implications for the rest of the world, especially the United States.

Even for the United States, the least vulnerable and most profligate of the oil-importing countries, the Arab embargo and oil price rise brought home a much needed lesson. In Henry Kissinger's words, "One embargo— and one economic crisis—should be enough to underline the implications of dependency." (Address before the National Press Club, Washington, D.C., February 3, 1975.) Americans should, therefore, have little difficulty in understanding Japan's far greater vulnerability and dependency and how imperative it must be for Japan to try to alleviate the oil threat. Americans, too, cannot but remain very alert to any significant change in the external position and economic activities of Japan, long regarded as the linchpin of U.S. policy in Asia and an indispensable factor in the global balancing scheme of America's new geopolitics.

Several related but different aspects of Japanese policy will be explored. In the first place, the manner in which Japan, when confronted with the Arab threat, shifted its foreign policy posture and redoubled its effort to resolve the country's oil supply and payments problems presents an excellent demonstration of alacrity and single-mindedness on the part of Japanese policymakers and business firms. At the same time, the critic might argue, the Japanese response also betrayed overanxiety and, given the benefit of hindsight, could perhaps be regarded as an overreaction. Thus, Japan's response to the 1973–74 oil crisis offers us a most interesting sample of Japanese political and economic behavior which should be given appropriate consideration in future evaluations of Japanese policy in crisis.

Second, within the framework of its overall energy policy, Japan's effort to safeguard its oil supply and to minimize the adverse income, employment, price, and balance of payments effects of the potential oil shortage and actual price increase boils down to a program of (1) diversification,

(2) increase in the proportion of oil supply under Japanese control, viz., the so-called "autonomous supply," (3) expansion of foreign exchange receipts through a renewed export drive, and, for certain periods, (4) monetary restraint and direct controls of oil consumption. Since Japan has undertaken these measures—which carry strong connotations of economic nationalism—and continues to undertake some of them when other nations must do the same, and since greater Japanese control of the oil Japan needs is also likely to run counter to the objectives of the nationalistic economic policies of the oil-producing countries, potential conflict could conceivably arise as a result of clashes of economic interests. Such contingencies are likely to involve Japan, the United States—and, to a lesser extent, other oil-consuming and importing countries—and the oil-producing, less developed economies. Consequently, what we propose to discuss here is a particular case or prototype of economic nationalism. What is true of oil can, of course, be equally true of other raw materials. The oil case presents a prime example, however, because of the nearly universal demand for oil by many industries and many nations and because of the dramatic fashion in which the oil crisis was precipitated.

In the third place, if, in the circumstances, Japan's economic policies and business conduct, aggravated by those of other nations, are not conducive to the convergence of different national economic interests, an inevitably adverse impact on other aspects of Japan's foreign relations may be expected. In particular, one's thought turns immediately to the U.S.–Japan security treaty, the successive shocks this bilateral relationship has already undergone since 1971, and the somewhat belated recognition in the United States of the need for repair and rebuilding of the relationship. Above all, we are reminded especially of the difficulty all nations in general must face when they try to reestablish national credibility and mutual trust as against the far greater ease with which credibility and trust can be undermined and destroyed. If U.S. policy continues to regard Japan as its most important partner in the Pacific and if Japan continues to regard the United States as an indispensable friend and ally, then the added obstacles which economic nationalism could erect must be minimized. Better understanding of the issues involved by the public, including the business community, and by policymakers of both countries is a pre-requisite of intelligent policy. Hopefully the present study will contribute a little to this much needed understanding.

As we shall attempt to show, in the sense of secure national control of the greater part of Japan's oil needs, the objective of real independence in oil supply may turn out to be illusory after all. At any rate, the extent to which this objective can be attained in the next few years cannot but be strictly limited. Yet it is precisely during this period that other aspects of international economic relationships—including those between the United States and Japan, and the broader conditions of the strategic balance and political relations among the United States, the Soviet Union, the People's Republic of China (PRC), and Japan—can and probably will change dramatically. It is possible, therefore, that Japan's redoubled efforts during this period in order to reach a long-term objective that may not be really attainable could well produce effects that are quite undesirable even from a narrowly nationalistic point of view. If interdependence rather than independence is what the future holds, would it not be better that national policies anticipate this condition as a reality? In this respect, while the present study deals with a single case and the external economic policy of Japan in particular, there are obviously far wider implications.

In the long run, a fundamental factor in Japan's oil policy may consist of a restructuring of the national economy in favor of (1) less energy-consuming products, (2) less energy-consuming technologies, and (3) greater substitution of oil by other forms of energy that can be more readily produced at home and/or imported from cheaper and more numerous and secure foreign sources. In their *Economic* and *Trade White Papers* of 1974, both the Japanese Economic Planning Agency and MITI are clearly thinking in these terms. The Energy Council's recommendation in 1975 on long-term energy policy reiterates the same themes. We shall touch on this general framework in Chapter 1. We do not propose to discuss in this study the details of these long-term policies or even how realistic they are in the light of past experience. Neither time nor space would permit us to broaden our inquiry beyond its present scope. More important to us here is the proposition that errors committed in the short run could totally alter parameters of long-term planning. In limiting our treatment to the Japanese effort to achieve diversification and a higher degree of national control of the nation's oil supply, we shall really be looking at some of the constraints of Japan's future policy and, more than just incidentally, of future U.S.–Japan relations.

Realization of the immediate relationship between the external policy of an oil-consuming nation and its need to secure oil, as manifested in Secretary Kissinger's November 1974 Chicago speech on energy, is not enough. As we shall see below, certain long-term effects of national economic policies on the pattern of foreign investments, the impetus of competition in foreign trade, and the interaction of the economic policies of Japan, the United States, and other developed and developing nations may be even more intractable. How we shall take these potential effects into account may present an even greater future challenge to policymakers on both sides of the Pacific. How ready are any of us to face this challenge with understanding beyond the usual platitudes?

REFERENCE CONVERSION TABLE

1 barrel of 42 U.S. gallons is equivalent to 158.99 liters.

1 kiloliter equals 6.2897 barrels.

1 kiloliter of crude oil averages 0.8 to 0.9 metric ton.

1 metric ton of crude oil has a volume of 1.1 to 1.25 kiloliters.

1 liter crude oil has an energy content averaging 9,400 kilocalories (Kcal).

In energy content, 1 kg. of LNG (liquid natural gas) is equivalent to 1.4149 liters of crude oil.

1 KWH is equivalent to 0.26 liter of crude oil.

(For additional conversion rates used in Japanese energy statistics, see *Comprehensive Energy Statistics* [1974], p. 371 and front of volume.)

BACKGROUND

A Chronicle of the 1973–74 Crisis

The oil ministers of the member states of the Organization of Arab Petroleum Exporting Countries (OAPEC) met in Kuwait on October 17, 1973,
and resolved to unsheath their oil weapon in order to induce the United
States and other large oil-consuming countries to change the latter's generally pro-Israeli attitude in the Arab-Israeli war. They decided at the
meeting

> that each Arab oil-exporting country immediately cut its oil production by
> a recurrent monthly rate of no less than 5 percent to be initially counted on
> the virtual production of September and thenceforth on the last production
> figure until such a time as the international community compels Israel to
> relinquish [the occupied Arab] territories or until the production of every
> individual country reaches the point where its economy does not permit of
> any further reduction without detriment to its national and Arab obligations.[1]

The same resolution mentioned specifically that the cuts applied to individual consuming countries "may well be aggravated proportionately with
their support to and cooperation with Israel." Acting on this principle, the
oil ministers recommended a total embargo of oil supply to the United
States. On the other hand, countries that supported the Arab cause or took
measures against Israel would "continue to receive the same oil supplies"
as before.

This general threat to reduce oil supply to all "Israel supporters" was
quickly followed by further efforts to make the threat more specific. On
October 19, 1973, ten Arab envoys headed by the Ambassador of Saudi
Arabia, which is Japan's main source of oil, called on Foreign Minister
Masayoshi Ohira, jointly requesting Tokyo's support for the Arab position
in the Middle Eastern war. Ohira responded by expressing his hope for an
early settlement of the war, stating that Japan would continue to uphold
UN Resolution 242, which was adopted by the Security Council in 1967
and called for the withdrawal of Israeli forces from territories occupied
during that earlier war. In a note from Vice Foreign Minister Hogen to

the Saudi Arabian Ambassador, this relatively mild general Japanese position was reaffirmed a week later. The note restated Japan's opposition to
territorial acquisition by force as well as its strong support of the 1967 UN
resolution.

OAPEC's threat of an oil embargo came upon the heels of a series of
oil price increases since 1971. For instance, on February 15, 1971, the
posted price of Arabian Light crude oil, a benchmark or reference point
used in pricing other crudes, was raised to $2.180 a barrel from $1.800, the
price announced on August 31, 1970. Subsequent increases of the posted
price were announced on June 1, 1971 ($2.285), January 20, 1972 ($2.479),
January 1, 1973 ($2.591), April 1, 1973 ($2.742), and August 1, 1973 ($3.066).
On October 1, 1973, shortly before the Kuwait meeting, the posted price
of Arabian Light was $3.011 a barrel. On October 16, in conjunction with
the ministers' meeting, the posted price was raised to $5.119 a barrel. (See
Table 1 for changes in the posted price and Japan's import prices.) Because
of the complex formula used in computing actual price from the posted
price, which determines the producing countries' "take," and a certain
degree of flexibility in pricing retained by the oil companies, the actual
import price may not have risen immediately to the same degree. An Associated Press report from Kuwait mentioned an increase of 21 percent for
the standard Arabian Light from $3.011 to $3.650 a barrel.[2] According to
the Japanese customs, the average CIF price of imported oil in October
1973 was $3.46 a barrel (5,778 yen a kiloliter). It rose to $4.09 a barrel
(6,984 yen a kiloliter) in November; in December it was $5.02 (8,840 yen).

The Arabian Oil Company, largest Japanese-owned oil-producing firm
overseas, had an average daily output of 420,000 barrels of crude oil during
September 1973. The company was ordered by the Saudi government to
reduce its production by 10 percent—which was twice the 5 percent uni-
form rate—and to hold future production at the reduced level until the
end of November. The government intimated that consideration would be
given to further measures at that time.[3] Simultaneously, Japanese refineries
supplied by British Petroleum, EXXON, and Gulf were notified by these
major international companies that crude deliveries to them would be
reduced henceforth. Other "majors," such as Shell, Mobil, and Caltex,
followed suit.[4] This chain reaction was transmitted further by the Japanese
refineries which immediately advised their clients that supplies of petroleum products would in turn have to be cut back.[5] The Arabian Oil Company reportedly was taken by surprise since the OAPEC countries had
indicated in their October 17 resolution that favorable treatment would
be meted out to friendly countries. The Japanese had apparently assumed
that Japan would be regarded as a "friendly" country. This assumption
now appeared to be mistaken.

It will be noted that these concrete steps took place during the last part of October, more or less coinciding with the note from the Japanese Foreign Office. It would appear that Japan's diplomatic gestures were deemed insufficient by the Arab states. Since the major target of Arab policy was the United States, the principal supporter of Israel, OAPEC's tactic was to build up enough diplomatic pressure on the United States from America's allies. Japan, like some countries in Western Europe, was singularly vulnerable and could not, therefore, be let off too easily.

The screw was tightened at the beginning of November. At a second meeting of the Arab oil ministers in Kuwait on November 4–5, 1973, the decision was made to reduce oil production in each Arab country that was party to the October 17 decision by 25 percent of the country's September production. Furthermore, another "reduction amounting to 5 percent of the November output will follow in December provided that such reduction shall not affect the share that any friendly state was importing from any Arab-exporting country during the first 9 months of 1972."[6] Thus the oil-exporting countries were threatening to reduce production further, placing the onus of the reduction entirely on the "unfriendly states." This threat soon proved effective.

On November 6 the foreign ministers of the European Economic Community (EEC) issued a joint resolution that in effect bowed to Arab demands. On the same day a statement reiterating the position previously announced by the Foreign Ministry was issued by the chief secretary of the Japanese Cabinet, Susumu Nikaido. This was followed by discussions between the Japanese prime minister and the visiting U.S. secretary of state. In the course of these discussions Prime Minister Tanaka reportedly stated on November 14 that in the prevailing circumstances Japan was compelled to pursue an independent diplomatic policy vis-à-vis the Arab nations. The new position of the Japanese government was then announced on November 22 in a second statement by Nikaido. The statement made three points: (1) Israel should withdraw from the territories it had occupied since the Six-Day War of 1967. (2) Japan subscribed to the Arab position that the right of self-determination of the Palestinians should be a precondition of any peaceful solution of the Middle Eastern problem. (3) The Japanese government would reconsider its policy toward Israel in the event of the latter's refusal to accept these preconditions. At the same time, the Japanese government also announced its plans to send Takeo Miki, then deputy prime minister (prime minister at this writing), on a goodwill tour of the Arab countries during December.

Following the open concession of the nine EEC governments, the Arab oil ministers announced on November 18, 1973, "the suspension of the 5 percent cut in oil production in December in relation to the countries of the European Economic Community except Holland, provided that the decrease continues afterward for all nonexcluded countries at the rate of 5 percent for January compared to the December production level."[7] This was followed by a statement of the Arab oil ministers on December 8, 1973, lifting the embargo conditionally. Then at a meeting in Kuwait on December 24–25, 1973, the Arab ministers took note "of the change in Japan's policy towards the Arab cause which has been demonstrated by various means including the visit of the Japanese Deputy Prime Minister . . . , and . . . of the deteriorating economic situation in Japan." They therefore "decided to accord Japan special treatment which would not subject it to the full extent of the across-the-board cutback measures, out of a desire to protect the Japanese economy and in the hope that the Japanese government, in appreciation of this stand, will continue to adopt just and fair positions vis-à-vis the Arab cause."[8]

This sequence of events suggests rather strongly that Tokyo issued its statement of November 22 after the Japanese government had witnessed the relatively favorable response of OAPEC on November 18 to the EEC concession of November 6. The discussion between Tanaka and Kissinger on November 14 may also have contributed to the Japanese decision, as it were, to surrender openly. The November 22 statement was a landmark in Japan's "resource diplomacy."

The second round in Japan's resource diplomacy consisted of a series of foreign tours undertaken by high officials to explain the Japanese position and to convey goodwill to the oil-producing countries. From December 10 to 28, 1973, Takeo Miki visited Saudi Arabia, Egypt, Abu Dhabi, Kuwait, Qatar, Syria, Iraq, and Iran. From January 7 to 17, 1974, Prime Minister Tanaka personally visited the Philippines, Thailand, Singapore, Malaysia, and Indonesia, all members of the Association of Southeast Asian Nations (ASEAN). Overlapping with Tanaka's Southeast Asia tour the already well-traveled Mr. Miki went to the United States, presumably to seek U.S. understanding of Japan's quandary. In the meantime, Yasuhiro Nakasone, head of the Ministry of International Trade and Industry (MITI), again called on Iran and Iraq. Also included in Nakasone's itinerary between January 7 and 18 were the United Kingdom and Bulgaria. Finally, those Arab countries that were missed by both Miki and Nakasone were visited by a former director general of the Economic Planning Agency and onetime foreign minister, Zentaro Kosaka. Kosaka's itinerary included Morocco, Algeria, Tunisia, Libya, Jordan, the Sudan, Lebanon, and Yemen. This very extensive series of visits took place on January 15–February 2. Thus, by the end of January 1974, three and one-half months after the onset of the October crisis, Japan had completed laying the foundation of its new resource diplomacy aimed at securing oil supply.

The Japanese officials visiting the Arab countries came with promises of gifts. During Miki's visit to Egypt, for instance, he promised $280 million of Japanese credit for both the first and second stages of a project to dredge and widen the Suez Canal, reportedly covering 70 percent of the estimated total cost.[9] Miki also agreed with Saudi Arabia on a possible economic and technological cooperation arrangement for the construction of chemical, oil-refining, steel, and copper-mining plants in the Arab country. During Tanaka's Southeast Asian visit, Japan promised Malaysia a third yen loan amounting to $100 million. Indonesia was promised a $200 million loan for the development of liquid natural gas in North Sumatra and East Kalimantan, as well as all-out Japanese support of Indonesia's Asahan power project for aluminum refining and the construction of oil-storage facilities on Lombok. Nakasone, in turn, promised $1 billion of credit to Iran for the construction of a 500,000 barrel-a-day joint Japanese-Iranian refinery, as well as another billion dollars of credit to Iraq for the construction of a 200,000 barrel-a-day joint refinery and an LPG (liquified petroleum gas) plant with an annual capacity of 1.8 million tons. In addition, Nakasone reportedly also extended to Iraq $350 million in credit for a petrochemical complex and $150 million for a cement plant. The quid pro quo of these Japanese offers was Iraq's agreement to supply a total of 160 million tons of crude oil, LPG, and other products during a ten-year period beginning from 1974. The cumulative amount of oil and LPG to be supplied could be boosted to 200 million tons. Other Japanese pledges to still other countries included those made by Kosaka to Algeria—$40 million of credit for refinery construction, oil and gas exploration, a new communications network and industrial development; $10 million of credit to Morocco and Jordan for a communications network; and $10 million to the Sudan for a fertilizer plant. One Japanese estimate has put the total amount of credits thus pledged in Japan's resource diplomacy in the critical winter of 1973–74 at $3.3 billion.[10]

The Internal Scene

On the domestic scene MITI maintained through most of October 1973 that there would be no need to control domestic energy consumption. By November 16, however, when Japan was explaining its energy woes to the U.S. secretary of state, certain "principles of countermeasures" to meet the oil emergency finally won cabinet approval. An energy-saving movement was initiated, and measures for reducing electricity and oil consumption in government offices were introduced. All these measures were supposed to be coordinated and implemented by a special headquarters under the direct command of the prime minister. The first countermeasures were either general in nature or specifically related to the public sector. They were followed a few days later by a separate administrative guidance dealing with the principles of saving oil and electric power in the private sector. Then, on November 30, an Oil Supply-Demand Adjustment Bill was approved by the cabinet along with another bill known as the National Livelihood Stabilization Emergency Measures. The two bills won passage in the Diet on December 25, 1973, coinciding, therefore, with the official inclusion of Japan by OAPEC among the nations friendly to the Arabs.

This flurry of measures aiming at the reduction, as well as administrative redirection, of the demand for oil occurred during the same period when the Arab threat of an oil cutback appeared most serious. As a part of the basic principles underlying the emergency countermeasures of November 16, private automobile driving was to be reduced on a voluntary basis. A ten percent reduction in the use of oil and electric power by business corporations was to be implemented during December. A five-day work-week was also advocated in order to save energy. Finally, gas stations were to be closed during holidays on a voluntary basis.

The administrative guidance for the private sector issued on November 19 reiterated the same general idea. Industrial users in such industries as iron and steel, automobile manufacturing, heavy machinery, electric appliances, petroleum, chemicals, automobile tires, synthetic fibers, aluminum and other nonferrous metals refining, cement, sheet glass, and paper and pulp were all strongly urged to save oil. The Oil Supply-Demand Adjustment Law first approved by the cabinet on November 30 also gave the government standby power to allocate and ration petroleum products and to restrict their use in the event of a really serious shortage. The accompanying National Livelihood Stabilization Law was intended to give the government additional emergency powers to control demand and supply in general, together with commodity prices.

When the initial energy saving program went into effect on November 16, the government called for a voluntary 10-percent reduction of oil and power consumption through the end of 1973. A second announcement in mid-December indicated that the reduction would be raised to 20 percent beginning in January 1974. However, with the designation of Japan by OAPEC as a "friendly nation" on December 25, the Japanese government decided to relax the proposed cutback which it seemed most reluctant to enforce in spite of the apparent flurry of legislative and administrative activities. On January 11, 1974, it reduced the proposed cutback of both electricity and oil consumption, originally scheduled to begin on January 16, 1974, to a maximum of 15 percent instead of the 20 percent proposed barely a month earlier.

With the conclusion of the tours of oil-producing countries by cabinet-level Japanese officials in January, relaxation of consumption control continued during the following months. A government decision was made on February 26, 1974, to relax the degree of industrial consumption of energy by 5 percentage points. Among the industries affected were foodstuffs, textiles, paper and pulp, chemicals, iron and steel, other metals, and machinery products. For certain industries the rate of oil reduction was changed from 13 to 8 percent and that of power reduction from 15 to 10 percent. For certain other industries such as the manufacture of detergents for household use, dairy products, paper products, etc., previous orders to reduce consumption either at 5 percent below the level of the corresponding month in 1973 or at no higher than the level of the month a year ago were completely rescinded. Closing of service stations on Sundays and holidays was continued, but neon lights and electrical advertising signs using less than 10 kilowatts of electricity were now allowed between six and nine in the evening. This process of gradual but progressive relaxation culminated finally in the announcement by Nakasone at the end of August 1974 that the energy emergency was over.[11]

An Emergent Long-term Energy Policy

Not quite a year later, the outline of Japan's long-range energy policy, drafted with the full benefit of more than a year's experience after the oil crisis, was revealed in a set of "recommendations" prepared by Japan's Energy Advisory Council.[12] Using FY (fiscal year) 1985 as a target year, the council's recommendations consist of a series of measures that aim at matching Japan's demand for energy to the anticipated supply. The latter is derived by first forecasting the world's supply and demand of energy outside the communist countries as well as available import from the communist countries (estimated at 150 as against 50 million kiloliters in oil equivalent in FY 1973) and a special world oil-flow model is then constructed. On the basis of these projections it is further assumed that certain energy imports by Japan, including 485 million kiloliters of crude oil and petroleum products, would be feasible. The detailed distribution of primary energy—both indigenous and imported—by source is given in Appendix Table A.

On the demand side, total energy demand is to be restrained by conservation and technological change. Energy-saving measures are given a potential success of reducing energy consumption by over 9 percent.

As for the demand for oil, this is to be done by shifting to other energy sources—above all to nuclear energy—within the next decade. Furthermore, there would be an increase in imported liquid natural gas (LNG) in lieu of crude oil, not the least for environmental reasons. If all measures are fully on target, imported oil and petroleum products would account for 63.3 percent of total energy supply in FY 1985 instead of the 77.4 percent in FY 1973.

Even then, however, this would still mean a more than 50 percent increase in import above the 318 million kiloliters (crude and products) imported in FY 1973. This supply, the council's recommendations stress, must be made *secure* and *stable*. To do so, Japan's resource diplomacy must be fully implemented and its recommendations are, in effect, two-fold: first, international cooperation (a) through membership in the International Energy Agency (IEA) in order to safeguard emergency-sharing and stockpile-building, (b) through financial participation in international development schemes, especially in oil-producing countries and in the light of the reduced power and altered role of the major international oil companies, and (c) diplomatic approaches aimed at ameliorating possible confrontation between oil-consuming and oil-producing countries; second, national programs to cultivate bilateral relations between Japan and individual oil-producing countries.

The recommendations also mention explicitly Japan's need to pay heed to its special relations with certain countries such as the United States. It should be noted, however, that in the context of these recommendations "international cooperation" is not identical with cooperation among the oil-consuming nations. From the point of view of the consuming countries only, the line between international cooperation and national programs in Japan's long-range energy policy—still largely an oil policy—becomes seriously blurred. It is fair to say, however, that this long-range policy of 1975 attempts to keep all options open. If international cooperation among the consuming nations and/or between consuming and producing nations is not forthcoming or fails to offer real and timely solutions, Japan will not have wasted its time in relying entirely on one set of alternatives. Some progress will have been made already in developing national solutions.

2

VULNERABILITY:
THEORY AND FACT

A Redistribution of the World's Output and Wealth

If we look beyond the short-lived embargo of winter 1973–74, the by far more lasting effect of OPEC's (Organization of Petroleum Exporting Countries) newfound solidarity is the dramatic increase in the price of oil. If we divide countries in the world into oil exporters and importers, disregarding those which are neither, this sharp price increase is essentially a bid by the exporters to redistribute the world's output—and existing assets—in their favor. The exporters acting as a group now want more of the world's output for their oil. To the extent they cannot do without oil, the importers must now offer more for what they import. If what they offer in return through current export is not enough—partly because the oil exporters do not buy enough—they must go into debt vis-à-vis the exporters or transfer to the latter ownership in productive assets which will yield future income for the exporters. There is, however, no escaping the fact that such a redistribution of income and wealth will make the oil exporters richer. The same process will also make the importers poorer unless through increase in production and productivity they can compensate for the larger transfer of goods to be made to the oil exporters through extra production. It is just possible that this increment in economic growth over and above what would otherwise have occurred had the price of oil not risen could be brought about through new capital investment in the oil-importing countries made possible by "oil dollars" invested by the exporters. If there are unemployed resources in the oil-importing countries and/or if productivity can increase rapidly, it is even possible for these countries to increase capital formation, financed by the inflow of oil dollars, without curtailing either their own consumption or investment. In this case, the bid of the oil exporters for a larger share of the world's output is met by an expansion of that output.[1] Of course, to say it is "just possible" that all this may happen is to imply that it may not—or not very smoothly—because (1) owners of the oil dollars may not wish to invest in some of the oil-importing countries or do so in sufficient amounts, (2) the investment process will have a time lag, and (3) the capital and oil-importing countries may not be able or willing immediately to absorb the oil-dollar inflow.

However, the world is not divided into a union of oil importers acting as a single entity facing a cartel of oil exporters. Nor is there an exchange through barter in which oil cannot be exported without a corresponding export on the part of the oil-importing countries or investment in real goods by the oil-exporting countries in the importing countries. Rather the exporters first receive payment in money. Hence they have to decide what currency (currencies) they will accept and up to what amount. They must further decide what to buy with this money through imports or foreign investment, and in each case they must decide on what and where. For the individual oil-importing country, too, its first worry is whether it can pay for the higher-priced oil in a manner acceptable to the sellers if it cannot find another source of supply or a substitute. It must next worry about whether the oil-exporting countries' decisions will enable it to make the necessary adjustment which may or may not lead to a viable condition, not to mention a faster pace of economic growth.

Let us examine this adjustment process more closely as it may appear to an individual oil-importing country.

Impact on income, employment, and price.   One major impact of a sudden decrease in the supply of an important fuel and industrial raw material such as oil on a given country consists of the contraction of income and employment that must follow if substitutes cannot be quickly found and/or if an equivalent reduction on the demand side, limited to imports alone, cannot be brought about readily. Such a reduction in supply will ultimately be reflected in a price increase of the material in short supply, even though this could be suppressed for a while through government control. More likely, the decrease in supply will be accompanied by a substantial price increase from the very beginning. Alternatively, a substantial price increase could be enforced by a monopoly without any artificial withholding of supply. In all these cases of price increase, the country must also be prepared to cope with the resultant changes in both absolute and relative commodity prices.

The change in relative prices between oil and other energy products and industrial raw materials will tend to bring about a reallocation of resources among end products, as well as among different inputs, along with appropriate changes in technology. These effects are essentially similar to those brought about by a purely quantitative restriction of supply that would eventually also lead to shifts of relative prices. However, in the real world, these adjustments are unlikely to take place very quickly.

In addition, what will happen to the economy following a large immediate increase in the price of oil resembles the process of a cost-push inflation which could be brought about by a sharp rise of money wages and is generally discussed in that context. Like a wage increase that cannot be readily absorbed in the profit margin, an increase in oil price would lead to an increase in the prices of products using oil that producers will want to charge. The immediate effect would be reduced sales and a smaller output of these products, thus lowering the quantity of oil demanded and, simultaneously, the level of employment and demand for other inputs, followed by the usual "multiplier" effects on aggregate income and expenditure.

If the sequence of events was originally initiated by an increase in money wages, economic theory tells us that under certain simplifying assumptions the following developments could take place. First, the rise of prices would tend to reduce real wages, thus offsetting the effect of the initial increase in money wages. However, such a general price increase would not be sustainable unless there is an increase in the supply of money. This is the familiar argument that a cost-push inflation cannot be sustained without a demand-pull inflation. The argument assumes the absence of imperfect competition under which producers may keep prices high, or even raise prices, in the face of declining sales. Also it assumes that people will not increase the rate at which they will spend money, thus sustaining sales at higher prices.

Second, in the absence of economy-wide monopolies and monopolistic labor unions, the unemployed labor and industrial capacity due to the production cutback would build up pressure for a reduction of prices and money wages. This, of course, again assumes perfect competition in both the commodity and labor markets, as well as mobility and a high degree of homogeneity of labor and of productive facilities. It also is a long-run rather than short-run argument.

In the case of an increase in the price of oil brought about by a foreign producers' cartel, the initial rise of commodity prices in importing countries would of course have the effect of offsetting some of the gains in the terms of trade previously acquired by the oil cartel through its oil price increase. Furthermore, since in some oil-importing countries unemployment and contraction of output will set up strong political pressure on the governments to resort to monetary as well as fiscally expansionary measures, instead of monetary restraint accompanied at most by selective measures to stimulate employment and output without generating inflationary pressures on prices, there is a good chance that an expansionary process will be forthcoming to sustain and even to augment the original cost-push inflation. Besides, as mentioned above, prices may be kept up or even rise further because of imperfect competition in the commodity markets. Also, lack of mobility and imperfect competition in the labor market will slow down the reemployment of the unemployed and prevent the reduction of money wages. Under these conditions, the oil cartel may well be tempted to call for another round of price increase. (A good example of such a development is the argument attributed to the shah of Iran that price inflation in the oil-importing industrial countries after October 1973 should be compensated by another increase in oil price. This he tried to promote in 1975, which saw a *nominal* increase of 10 percent in oil price in October.)

Third, in the oil case, the unemployed workers in the oil-importing country could not really be reemployed through a decline of money wages, even if that were possible given the powers of trade unions, if we assume *totally inflexible* input coefficients accompanied by a cartel-imposed quantitative restriction in oil supply—the extreme case being a total embargo—*and* if the structure of demand does not change in favor of products requiring less oil to produce. Hence, given inflexibility in the demand structure and very slow changes in production methods and the underlying technology, the short-term impact of a sharp oil price increase can be quite painful, and it may take a long time before the necessary adjustments can be completed. It may turn out that some countries will be unable to withstand the political and social unheavals that prolonged economic recession would entail, quite apart from the severe economic effects themselves. In the eyes of Japanese leaders, their country is especially vulnerable in this respect due to Japan's overdependence on Arab oil, a high degree of inflexibility in the structure of domestic oil demand, and the trend of the governing Liberal Democratic Party's (LDP) declining hold on the electorate. These considerations may have been responsible for the government's reluctance to impose controls immediately even as its anxiety mounted in November–December 1973.

Impact on the balance of international payments. Even if the quantitative restriction has been removed following the lifting of the oil embargo, and even if domestic price increases of many commodities have partly offset the oil cost increase for their producers and production and employment levels have been substantially restored, there can still be a very serious international payments problem for individual oil-importing countries. To see this problem in its proper perspective, one has to realize that unless the oil price increase by the producers' cartel is matched by a corrresponding increase of equal proportions in the prices of all the commodities the cartel buys from the oil-importing countries, the terms of trade will have changed in favor of the oil-exporting countries and a transfer of real wealth will have occurred. This is true especially if the oil-exporting countries do not export other goods so that the terms of trade of their non-oil exports could not deteriorate in relation to their imports from the oil-importing countries. In varying degrees the individual oil-importing countries must now offer more than before in return for the oil they import, and some of the oil-importing countries will have a harder time than others.

Let us now consider the case of an individual country such as Japan. If it has to pay a great deal more than before for the oil it imports, how might this larger payment be effected? A combination of the following alternatives comes to mind: (1) by exporting more to some or all of the oil-exporting countries during the same period in which oil is imported; (2) by exporting more now in return for the promise of oil in future periods —that is, exchanging present export in the form of investment for future oil import, perhaps at a lower price and/or under a long-term contract; (3) by exporting more to other oil-importing countries in order to acquire means of payment, e.g., convertible foreign currencies that are acceptable to the countries to which the payment for oil must be made. These three methods have one point in common. They all rely on an expansion of current exports by this particular country.

In addition to using current exports to pay for the individual country's larger oil bill, still another alternative is to pay with wealth accumulated in the past. This can take the form either of an immediate transfer of assets to the oil-exporting countries or of contracting new indebtedness to the latter, to be repaid later. If borrowing, both short- and long-term, has its limits, then transfer of assets to the oil-exporting countries must be made. This, together with borrowing from the oil-exporting countries, is an important aspect of what is popularly described as the "recycling of the oil dollar." What kind of assets would our particular country have that it would be willing to transfer to ownership by oil-exporting countries? What would the latter be willing to accept? On what terms would such transfers take place?

First, perhaps the simplest way is to transfer foreign exchange assets that the oil-importing country currently possesses. For instance, Japan could pay for the balance of its oil bill not covered by current exports with U.S. dollars which it has previously accumulated. This is like the third method mentioned in the preceding paragraph. Given an ample Japanese supply of the currency in question, continued use of this method depends upon how far the oil-exporting countries are willing to let their holdings of the particular third-country currency (or currencies) rise. After all, the asset acquired in this manner is only in the nature of a promissory note, acceptability of which by any one country is predicated upon its general acceptability by most other countries. The stability of the external value of the third-country currency; the stability of that country's domestic prices, which is closely related to the first criterion; and the acceptability to most countries of said currency as an international reserve currency, which is largely determined by the two preceding factors—these are the determining factors that will affect how our oil-importing country can use its third-country currency holding to pay its oil bill. Assuming that these conditions are fulfilled, in the long run—if the larger oil bill continues—the oil-importing country will be compelled to try to replenish its holding of the third country's currency. This means that the oil-importing country must resort to the third method mentioned above, which is to realize an export surplus in trading with the oil-importing third country or third countries. Taking Japan as the oil-importing country and the United States as the oil-importing third country, Japanese payment of their oil bill with dollars must be predicated upon the continued acceptability of the U.S. dollar as an international medium of exchange and reserve currency. While this condition does not preclude other third-country currencies or SDR (special drawing rights) from playing a similar role, the American dollar must possess these qualities itself. Continued Japanese use of the U.S. dollar for such payment purposes is possible only if Japan's dollar holding can be replenished through investment income, dollar credit from the United States, other capital inflows, and new Japanese export surpluses, primarily—though not necessarily entirely—in direct trading with the United States.

Even if the oil-exporting countries are willing to accept, in payment for their export surplus, third-country currencies from the oil-importing countries—or, in the case of an individual oil-importing country, the latter's own currency—there may still be a long-term problem. When these currency holdings become very large, their sudden conversion to other currencies or other assets could play havoc with the stability of the currencies in question and create major liquidity problems in the world. During 1974–75 the oil-rich nations were apparently prudent in this respect and no such contingency arose. However, from the point of view of international exchange stability, it might be safer if the large asset transfer does not result in piling up currency holdings in the hands of countries that are politically volatile.

A different form of assets that the oil-exporting countries might find acceptable in the asset transfer process consists of real assets rather than paper obligations of the governments or monetary authorities of third countries. Examples of such real assets are real estate, factories, mercantile establishments, banks and financial houses, and productive units in all the economic sectors of oil-importing countries. After all, acquisition of such productive assets would provide the oil-exporting countries with sources of future income. Since oil is a wasting or depleting asset that would ultimately be exhausted, sources of future income must be created while the oil lasts. Internal economic development and acquisition of foreign assets for future income are not only alternatives; they also complement each other. Future internal economic development in the oil-exporting countries will require imported supply which income from foreign assets owned by them can help finance. In addition, the current rate of effective internal economic development in some of the oil-exporting countries may be severely limited by such constraints as inadequate infrastructure of capital goods and absence of skilled labor, and competent technical, managerial, and government personnel. In fact, for the economically less-developed oil-exporting countries, it may not be practical—and it may be wasteful—to try to spend its export surplus from oil immediately on internal economic development beyond severe limits. In the case of some very small oil-producing countries, size alone may set a sharp limit to the absorptive capacity of capital investment for internal development. While material capital goods can be purchased instantly, the human capital in joint demand with capital equipment and other development expenditures can only be acquired slowly and through domestic capital formation, that is to say, through the country's educational and training system and cultural transformation in a broad sense.

Finally, some assets within the category of government consumption may also be in demand by the oil-exporting countries, such as air defense systems, fighter planes, and other modern weapons systems. This is true especially in the case of Middle Eastern and Southeast Asian oil producers who feel threatened either by their neighbors or by domestic dissidents and insurgents. For some of them their newfound wealth in high-priced oil may have contributed to the perception of a heightened threat.

However, transfers of ownership and of the implied economic benefits of ownership may result in transfers of control. From the point of view of the present owners, transfer of certain assets to foreign ownership may influence the direction and manner of use of these assets in the future. National security interests, economic nationalism, and downright xeno-phobic sentiments all come into play. Since the entire asset transfer process under discussion has been occasioned by actions of OPEC, the Arab countries in particular, future political use of assets under their control may appear more likely to others by virtue of their past and present use of the oil weapon than it would otherwise be. We need to remind ourselves at this point of the truism that mutual trust between men taken a long time to rebuild once it has been severely undermined. Some oil-importing nations may be especially reluctant to transfer ownership of certain assets to foreigners. Thus, Iran's acquisition of an interest in Krupp would not be regarded with the same equanimity in West Germany as that of, say, German toy factories. National attitudes also differ greatly. For example, while the United States might have no strong objection, at least outwardly and with some safeguards, to the introduction of an Iranian financial interest in Pan American Airways, it would be virtually inconceivable to this writer for Japan to sell even a small minority interest of Japan Airlines to any foreign country, e.g., Indonesia. Individual countries will wish to retain unquestioned national control over certain sectors of their econo- mies, even though these sectors may vary from case to case.

In the particular case of Japan, the assets-transfer issue concerns the types of investments, other than portfolio investment of a noncontrolling magnitude, that it will permit the oil-exporting countries to acquire in Japan either through new investment or through acquisition of existing enterprises. If, for example, Japan would not allow its own computer firms, hotel resorts, or other assets to pass to Arab ownership, then it may have to be ready to sell Japanese-owned hotels in Waikiki or Japanese-owned banks in California to Arab buyers should such asset transfers be neces- sary. Such a shift of the locale of transferable assets to third countries cannot be made unless the oil-exporting countries desire the assets in their third-country location and the third countries have no objection of their own. (We presume, of course, that such transfers cannot be made surrep- titiously because of their magnitude. This assumption may not always be true.) Again, for Japan, no such transfers of assets located in third coun- tries can be made in the long run unless new Japanese export surpluses can be created for the initial asset acquisition or replacement. Assuming mutual convertibility of third-country currencies with one another at rela- tively stable rates, Japan will not have to create an export surplus in trade with specific third countries. However, the future prospects of the indi- vidual national economies in the eyes of the oil-exporting countries may ultimately affect their choice regarding the location of the foreign assets acceptable to them.

In conclusion, if Japan cannot pay for its larger oil bill by increasing direct exports to the oil-producing countries, it will have to pay by indirectly exporting more to other countries. Otherwise, it would have to open its doors wider to ownership of Japanese domestic assets by the oil- producing countries, assuming that the latter are unwilling to extend credit to Japan indefinitely or to be satisfied with portfolio investments in uncontrolling interests. Since other oil-importing countries, such as Western Europe and the United States, must resolve similar problems induced by the high-priced oil, each may try to create an export surplus with the others in order to pay the oil-producing countries as a group what each cannot pay through direct exports and transfers of domestic assets to the latter. Using the United States for illustration, the first major question is whether the United States would be willing to permit Japan to continue to pile up dollar holdings in bilateral U.S.–Japan relations. Assuming an affirmative answer to the last question and that the newly earned Japanese dollar-holdings will be continuously transferred to the oil-producing countries, which will then spend the money on productive assets in the United States, the next major question is whether the United States will permit such real asset transfers indefinitely. For the United States, the magnitude of the asset transfer problem due to its own oil bill is magnified by the necessary asset transfers "for Japan's account." If we substitute another country for Japan and/or Western Europe for the United States, an even more complicated picture will emerge. The nature of economic conflicts inherent in such national approaches to the oil issue and in economic nationalism in general remains the same.

It should be quite clear by now that the existence of more than one oil-importing country suffering from the economic effects of the dual impact of the oil crisis—a brief period of moderate quantitative restriction and a prolonged period of high price—could lead to competitive and mutually incompatible national efforts to solve the individual countries' balance of payments problems. Interaction of these national economic developments and national policies, however, can be even more complex. One of the underlying reasons is disparity in the alertness of policymakers in different countries. Another is the unequal effort individual countries will make to help themselves and the varying degree of success each will have. Still another reason lies in differences in their basic economic conditions. A country like Japan that is highly sensitive to the physical availability of oil fears the severe impact on income and employment that an oil embargo could bring. Yet the oil price hike and the induced cost-push inflation could lead to a reduction of production that greatly reduces the amount of oil demanded.

If other countries have the same problem, then the reduction of real income and employment in the different countries will affect their respective exports through changes in import demand and export availability— again, of course, to varying degrees. Not only are the income elasticities of demand for imports of the different countries unequal, but their competitive strengths on each other's markets are different. Besides, many export commodities of the different oil-importing countries will also rise in price in varying degrees in the course of the general rise of prices, while some of their exports conceivably may even fall in price. Thus, it is entirely possible for an oil-importing country to have a fortunate combination of circumstances so that a significant reduction of the amount of oil demanded following a small reduction in overall output and employment is accompanied by a large increase in its total export value, in spite of a possible contraction of quantity, made possible by large price increases of its special exports, reinforced by highly competitive export marketing. Such a combination of circumstances could even temporarily lead to a surplus of oil supply over demand and a payments problem, as well as domestic price and output instability, seemingly under control. Actual developments in Japan in the post-oil-crisis year seemed to present exactly such a situation, as some observers might interpret. Yet a less fortunate or economically less effective country may have to experience simultaneously a much greater contraction of national output and employment, a serious international payments problem, and continued oil shortage. During 1974–75 many more nations seemed to be closer to the latter than the former position. Divergencies in the degree of success of different countries in adjusting to the new economic situation created by the oil price hike and the threatened quantitative cutback during the initial period will create new pressures for further adjustment later. A priori, it does not follow that the initial success of a particular country will necessarily be followed by continued success. This point and a more detailed review of developments in Japan in 1974–75 will be presented in the last section of this chapter.

Some Basic Facts Underlying Japan's
Perception of Special Vulnerability

Japan's special vulnerability to OAPEC's embargo threat and supply cutback can be easily understood if we bear in mind the following facts:

First, Japan has a negligible amount of domestic oil supply. Virtually all crude oil and some distillates are imported.

Second, in spite of a decline during the latter part of the 1960s, the Arab countries have continued to supply a very large share of total imports. The share would be even larger if we added Iran as the largest non-Arab supplier in the Middle East.

Third, Japan had a relatively small inventory in the home islands before the onset of the 1973 crisis.

Fourth, over the years oil has become an increasingly larger source of all energy products so that a cutback of imported oil supply would be tantamont to a large reduction of all energy supply.

Fifth, only a very small share of the country's supply of oil and petroleum products is used in direct personal consumption. Hence contraction of personal demand would not contribute significantly to that of overall demand for oil. A substantial overall contraction can come about only

TABLE 1

CRUDE OIL PRICE
1970–74

(dollars per barrel)

Japan's average CIF import price

Year		Average (all types)
1970		1.80
1971		2.18
1972		2.51
1973	January	2.58
	February	2.71
	March	2.76
	April	2.82
	May	3.02
	June	3.09
	July	3.18
	August	3.25
	September	3.31
	October	3.46
	November	4.09
	December	5.02
1974	January	6.06
	February	9.90
	March	10.51

Changes in posted price[a]

	1970	1971		1972	1973	
	Aug. 31	Feb. 15	June 1	Jan. 20	Jan. 1	Apr. 1
Arabian Light	1.800	2.180	2.285	2.497	2.591	2.742
Arabian Heavy	1.470	1.960	2.064	2.239	2.345	2.481
Iranian Light	1.790	2.170	2.274	2.467	2.579	2.729
Iranian Heavy	1.630	2.125	2.228	2.417	2.527	2.674
Sumatra Light	1.700	2.210	2.260	2.960	—	3.730
		(Apr. 1)	(Oct. 1)	(Apr. 1)		

	1973					1974
	Aug. 1	Oct. 1	Oct. 16	Nov. 1	Dec. 1	Jan. 1
Arabian Light	3.066	3.011	5.119	5.176	5.036	11.651
Arabian Heavy	2.775	2.725	4.633	4.684	4.557	11.441
Iranian Light	3.050	2.995	5.341	5.401	5.254	11.875
Iranian Heavy	2.989	2.936	4.991	5.046	5.006	11.635
Sumatra Light	—	4.750	—	6.000	—	10.800

SOURCE: *Seikiyu Kaihatsu Kankei Shiryo*, [Data on Oil Development] (Petroleum Producers Federation, 1974) pp. 8, 113.

[a] The "posted price" is a fictitious price originally used to calculate the royalty and tax payments an oil company has to pay the government of the oil-producing country. These payments, together with the production cost and the oil company's profit, would then constitute the real price. Hence varying the posted price would lead to change in the real price in the same direction if the other cost elements and ratios remain constant. In practice, however, they have not always remained constant. Where the oil produced is initially entirely or partly owned by the oil-producing country under various degrees of nationalization, the oil company must then "buy back" the oil for export. The posted price is often used in this case as a reference point for setting the "buy-back" price, which in 1974 was often at 93 percent of the posted price.

through an immediate, direct contraction of industrial production and employment, affecting many of Japan's leading export industries.

Sixth, up to 1973, Japan had become accustomed to sustained large annual increases of its GNP. A decline, or even a lowered rate of growth, would represent a severe blow, more so psychologically than economically.

Seventh, Japan had also become accustomed to large and rapidly rising exports so that any threat to the continuation of this trend would be a traumatic experience psychologically, even though the ratio of Japanese exports to GNP was not particularly high.

Lastly, a series of international economic shocks had already been sustained by Japan since 1971. Further blows to the country's payments position could undermine its competitive position in the world fundamentally. The above points can be substantiated by the following statistics:

In 1953, when the Korean War ended, petroleum and imported petroleum products constituted only 17.7 percent of the country's total primary energy supply (see Table 2). The Japanese economy was then still highly dependent on coal, which accounted for 52.8 percent of the total. The importance of oil, however, rose steadily, exceeding the 25 percent level in 1957–58, the 50 percent level in 1962–63, and, finally, the 75 percent level in 1972–73 on the eve of the Arab embargo threat.

As shown in Table 2, the increasing importance of oil as a source of energy means a growing dependence upon imported oil. During 1953, when oil and imported petroleum products constituted 17.7 percent of the country's total primary energy supply, the proportion between import and domestic supply was 17.1 percent to 0.6 percent of the total. In 1973, 77.4 percentage points of the 77.6 percent of total energy supply came from import, and only 0.2 percent point from domestic production. If we leave out petroleum products and look at crude oil alone, in 1973 imported crude amounted to about 99.7 percent of aggregate supply excluding inventory on hand (see Table 3).

The increase in Japan's dependence on imported oil over a long period was doubtless encouraged by the low price of oil; it also reflected changes in the Japanese economic structure in favor of oil-using products and technologies. This shift toward more energy and oil-using products can be seen in the following statistics. Thus, for each billion yen of the country's GNP measured at 1965 prices, the amount of total energy consumption rose from 41.4 billion kilocalories in 1961 to 44.8 billion in 1965, 49.4 billion in 1970, and 49.3 billion in 1973.[2]

A convenient index of the relationship between economic growth and the demand for energy—hence for oil—is the "income elasticity of demand for energy" constructed by dividing the percent rate of growth of energy consumed over a given period of time by the growth rate of the GNP

TABLE 2

Percentage Distribution of Japan's Primary Energy Supply by Product
1953–73

Fiscal Year	Hydropower	Nuclear Power	Coal Domestic	Coal Import	Coal Total	Imported Coke	Lignite	Petroleum and Imported Petroleum Products Domestic	Petroleum Import	Petroleum Total	Natural Gas	LNG	Charcoal	Fuel Wood	Total Import	Total Domestic	Total	Total in Billion Kcal	Percent Increase over Preceding Year
1953	19.7	—	46.8	6.0	52.8	—	1.2	0.6	17.1	17.7	0.2	—	2.8	5.6	23.1	76.9	100.0	533,660	—
1954	20.9	—	46.3	4.6	50.9	—	1.1	0.6	18.0	18.6	0.2	—	2.7	5.6	22.6	77.4	100.0	535,140	0.3
1955	21.2	—	44.8	4.4	49.2	—	1.0	0.6	19.6	20.2	0.4	—	2.6	5.4	24.0	76.0	100.0	560,160	4.7
1956	20.0	—	44.7	5.0	49.7	neg.	1.0	0.5	21.4	21.9	0.4	—	2.3	4.6	26.4	73.6	100.0	643,590	14.9
1957	19.0	—	42.7	6.5	49.2	neg.	0.9	0.5	23.6	24.1	0.6	—	2.2	4.0	30.1	69.9	100.0	729,520	13.4
1958	21.2	—	41.1	4.9	46.0	—	0.9	0.6	24.9	25.5	0.6	—	1.8	4.0	29.8	70.2	100.0	702,720	-3.7
1959	19.0	—	36.5	5.5	42.0	—	0.7	0.6	32.1	32.7	0.9	—	1.3	3.4	37.6	62.4	100.0	795,010	13.1
1960	15.3	—	34.4	7.1	41.5	—	0.6	0.6	37.1	37.7	1.0	—	1.1	2.8	44.2	55.8	100.0	937,490	17.9
1961	15.4	—	31.3	8.6	39.9	0.1	0.5	0.8	39.2	39.9	1.2	—	0.8	2.3	47.9	52.1	100.0	1,080,140	15.2
1962	13.4	—	28.7	7.3	36.0	neg.	0.4	0.8	45.3	46.6	1.3	—	0.6	2.2	52.6	47.4	100.0	1,138,950	5.4
1963	13.0	—	24.0	7.0	31.0	—	0.3	0.6	51.2	51.8	1.5	—	0.5	1.9	58.2	41.8	100.0	1,304,840	14.6
1964	11.6	—	21.8	7.4	29.2	—	0.2	0.5	55.2	55.7	1.4	—	0.3	1.6	62.6	37.4	100.0	1,461,170	12.0
1965	11.3	neg.	19.1	8.2	27.3	—	0.1	0.4	58.0	58.4	1.2	—	0.2	1.5	66.2	33.8	100.0	1,656,140	13.3
1966	10.7	0.1	17.4	8.8	26.2	neg.	0.1	0.4	60.0	60.4	1.1	—	0.2	1.2	68.7	31.3	100.0	1,825,520	10.1
1967	8.3	0.1	14.4	10.2	24.6	0.1	0.1	0.4	64.2	64.6	1.1	—	0.1	1.0	74.5	25.5	100.0	2,055,210	12.6
1968	7.8	0.1	12.4	11.2	23.6	neg.	0.1	0.4	66.1	66.5	1.0	—	0.1	0.8	77.4	22.6	100.0	2,347,780	14.2
1969	7.0	0.1	10.5	12.3	22.8	neg.	neg.	0.3	68.3	68.6	0.9	0.1	0.1	0.6	80.4	19.6	100.0	2,706,870	15.8
1970	6.3	0.4	8.1	12.6	20.7	neg.	neg.	0.3	70.5	70.8	0.9	0.4	neg.	0.5	83.5	16.5	100.0	3,104,680	14.7
1971	6.7	0.6	6.3	11.2	17.5	neg.	neg.	0.2	73.3	73.5	0.9	0.4	neg.	0.4	84.9	15.1	100.0	3,206,110	3.3
1972	6.3	0.7	5.3	11.3	16.6	—	neg.	0.2	74.7	74.9	0.8	0.4	neg.	0.3	86.4	13.6	100.0	3,443,380	7.4
1973	4.6	0.6	3.8	11.7	15.5	neg.	neg.	0.2	77.4	77.6	0.7	0.8	neg.	0.2	89.9	10.1	100.0	3,825,760	11.1

SOURCE: *Sogo Enerugii Tokei* (hereafter cited as *Comprehensive Energy Statistics*), edited by Agency of Natural Resources and Energy Directorate, Gen. Secretariat, Tokyo, 1974, pp. 176–89. Components may not add up to totals due to rounding.

TABLE 3

Japan's Supply and Demand of Crude Oil

1958–73

(in million Kl)

Fiscal Year	Beginning Inventory	SUPPLY Domestic Production	Import	Total	DEMAND Refining	Power Industry	City Gas	Other Industrial Use[a]	Total	Error + or –	Ending Inventory
1958	1.15	0.41	16.94	18.50	17.12	–	neg.	–	17.12	–0.19	1.19
1959	1.19	0.48	25.00	26.67	24.55	–	0.12	–	24.67	–0.14	1.86
1960	1.86	0.63	32.88	35.37	32.58	–	0.27	0.29	33.14	+0.08	2.31
1961	2.31	0.77	39.17	42.25	39.27	–	0.63	0.52	40.42	+0.37	2.20
1962	2.20	0.88	47.26	50.34	47.08	0.01	0.83	0.67	48.59	+0.72	2.47
1963	2.46	0.88	62.40	65.74	60.98	0.34	0.78	0.78	62.88	+0.69	3.56
1964	3.56	0.74	74.16	78.46	73.04	0.70	0.88	0.84	75.46	+0.58	3.58
1965	3.58	0.79	87.62	91.99	84.43	0.72	1.02	0.86	87.03	–0.08	4.88
1966	4.89	0.87	104.16	109.92	101.53	1.42	1.13	0.87	104.95	+0.05	5.02
1967	5.02	0.88	131.02	131.02	122.27	2.19	1.22	0.67	126.35	–0.16	4.51
1968	4.52	0.86	146.85	152.23	140.04	3.00	1.10	0.67	144.81	–0.11	7.31
1969	7.31	0.80	174.60	182.80	166.29	3.94	1.27	0.69	172.19	–0.76	9.85
1970	9.84	0.90	204.87	215.61	193.49	7.25	1.29	0.60	202.63	–1.67	11.31
1971	11.31	0.87	224.25	236.43	208.70	10.99	1.20	0.37	221.26	–1.78	13.39
1972	13.39	0.83	246.10	260.32	224.11	17.60	0.95	0.23	242.89	–0.71	16.72
1973	16.72	0.82	288.49	306.03	260.92	23.60	0.45	0.18	285.55	–1.27	19.60

SOURCE: *Comprehensive Energy Statistics, 1974*, 196–TG. See also *Nihon no Sekiyu Mondai* [Japan's Petroleum Problem], (MITI: Tokyo, December 1973), Table 2-1, p. 100.

[a] e.g., fertilizers

during the same period. The numerical value of this coefficient then gives us the percent increase in energy demanded for a 1.0 percent increase in GNP. It can be used in estimating demand for energy given certain changes in GNP, assuming no significant change in technology and relative demand among commodities that differ considerably in energy use. As is to be expected, Table 4 shows the steady rise of Japan's income elasticity

TABLE 4

GROWTH OF JAPAN'S GNP AND ENERGY CONSUMPTION
1958–72

Fiscal Year	Annual Growth Rate in Percentage (preceding year = 100)		Income Elasticity of Energy Consumption (II ÷ I)
	GNP at 1965 Prices (I)	Energy Consumption (II)	
1958	5.7	0.7	0.12
1959	11.7	14.9	1.27
1960	13.3	16.2	1.22
1961	14.4	14.4	1.00
1962	5.7	8.4	1.47
1963	12.9	14.0	1.09
1964	10.8	12.0	1.11
1965	5.4	9.1	1.69
1966	11.8	14.0	1.19
1967	13.4	14.5	1.08
1968	13.6	12.3	0.90
1969	12.4	17.0	1.37
1970	9.3	13.6	1.46
1971	5.7	4.7	0.82
1972	11.0	8.1	0.73
1955–60 average	9.1	10.5	1.15
1960–65 average	9.8	11.6	1.18
1965–70 average	12.1	14.3	1.18
1965–70:			
Average, U.S.	3.9	4.6	1.18
Average, U.K.	2.9	1.5	0.52
Average, France	5.8	5.7	0.98
Average, W. Germany	4.9	4.5	0.92
Average, Italy	5.5	9.8	1.78

SOURCE: *Sekiyu Kaihatsu Jiho* [Petroleum Development Journal], vol. 9, no. 23 (Petroleum Producers Federation [Sekiyu Kogyo Renmei], 1974), p. 102.

of energy consumption. The average for 1965–70 was 1.18, the same as a corresponding estimate for the United States, but considerably higher than that for West Germany, France, and the United Kingdom. (A small decline in the post-crisis period is inconclusive because of the brief period of time involved so far and possible asymmetry between changes in energy demand during a recession and demand changes in expansion.)

The preceding statistics point to the high sensitivity of Japan's energy demand to the level of production. *Mutatis mutandis*, if the supply of energy is arbitrarily reduced, the total supply of goods and services would have to be cut back, unless adjustments are made in technology and in the structure of product demand. Table 5 shows that a policy of making adjustments in the structure of demand cannot go very far by merely curtailing direct personal consumption of energy. Direct personal consumption probably accounted for no more than 25 percent of the total consumption of crude oil and petroleum products in 1972, the balance being taken up by nonpersonal consumption.[3]

TABLE 5

FINAL DEMAND FOR PETROLEUM AND PETROLEUM PRODUCTS BY SECTOR
FY 1972

	Amount of Energy (in billion Kcal.)	Percentage
1. Mining and industry		
a. Iron and steel	117,050	5.7
b. Other	521,520	25.5
c. Sub-total (a + b)	638,570	31.2
2. Energy production	152,660	7.4
3. Transportation[a]	382,810	18.7
4. Agriculture, forestry and marine products	70,300	3.4
5. Other personal consumption[b]	317,350	15.5
6. Total use as energy (1c + 2 + 3 + 4 + 5)	1,561,690	76.2
7. Non-energy uses (e.g., fertilizer production)	343,170	16.8
8. Total domestic demand (6 + 7)	1,904,860	93.0
9. Export[c]	143,900	7.0
10. Total	2,048,760	100.0

SOURCE: *Comprehensive Energy Statistics*, 1974, pp. 170–71, 209.
[a]About 55 percent in the form of gasoline for motor vehicles.
[b]Primarily heavy oil and jet fuel in roughly equal amounts.
[c]Primarily heavy oil, followed by jet fuel.

Japan's vulnerability to a reduction of oil import is further magnified by the large share represented by a few exporting countries. In 1972 the Arab countries (i.e., the Middle East and North Africa excluding Iran) were responsible for 43.4 percent of Japan's total crude oil import (see Table 6). The October 17 OAPEC communiqué called for an initial curtailment of 5 percent. Hypothetically, applying this figure to Japan's import, it would be equal to a 2.2 percent cutback of Japan's crude oil supply, or a 1.65 percent reduction of total primary energy.[4] Depending upon the

TABLE 6

DISTRIBUTION OF JAPAN'S CRUDE OIL IMPORT BY SOURCE

(percentage)

COUNTRY	1967	1972	1973
Middle East			
Saudi Arabia	18.2	16.7	19.9
Kuwait	17.5	8.9	8.3
Neutral Zone	14.1	8.4	5.3
Iran	35.9	37.3	31.0
Iraq	2.1	0.1	0.3
Qatar	0.3	—	0.1
Abu Dhabi	3.1	9.3	12.7
Subtotal	91.2	80.7	77.6
Indonesia			
Sumatra	6.2	2.7	13.5
Kalimantan	0.3	13.6	0.1
New Guinea	0.1	—	—
Malaysia			
Sabah	0.1	0.1	4.8
Subtotal	6.7	16.4	18.4
North America	0.1	—	—
Latin America			
Venezuela	0.4	0.2	0.2
Peru	—	—	—
Brazil	—	—	—
Argentina	—	—	—
Subtotal	0.5	0.3	0.2
Other			
Soviet Union	1.4	0.2	0.5
PRC	—	—	0.6
Rumania	0.2	—	—
Africa (Nigeria)	—	2.4	2.7
Egypt	—	—	—
Australia	—	—	—
Subtotal	1.6	2.6	3.8
Total	100.0	100.0	100.0

SOURCE: Comprehensive Energy Statistics, 1974, pp. 189–99.

income elasticity of energy consumption, in the absence of counter-measures GNP would have to decline accordingly. For instance, if elasticity is 1.18 (average of 1965–70), a 5 percent cutback of imported oil from the Arab countries would result in a 1.4 percent decrease in GNP. A 25 percent cutback—the full extent of the curtailment threatened— could then result in a 7 percent drop of the Japanese GNP, assuming as before the absence both of countermeasures and of other accentuating adverse effects. This again explains Japan's perception of vulnerability.

Apart from the concentration of the source of foreign oil in a few countries in the Middle East, a preponderant portion of this oil is imported by a small number of major international oil companies based in Western countries. The U.S.- and British-based companies alone were responsible for 66 percent of Japan's total crude oil import in FY 1973 (see Table 7). This particular combination of geographical and political grouping of Arab states as suppliers and of U.S.- and U.K.-based oil companies as importers presents a peculiar potential threat from the Japanese point of view. First, in the event of a general oil shortage in the world market, the foreign importers conceivably could channel oil away from Japan.[5] Second, as Japan sees it, the major oil companies conceivably could be directed by their own governments to do the same if the present cordial relationship between Japan and the West were once again replaced by mutual hostility

TABLE 7

JAPAN'S CRUDE OIL IMPORT BY NATIONALITY OF SUPPLIER

FY 1973

(percentage)

U.S. Companies	
Caltex, Exxon, Gulf, Mobil, Getty, Union, and other	50.7
British Companies	
SIPC and BP	15.4
Subtotal	66.1
Japanese Companies	8.5
Other (including DD and GG oil and oil supplied by firms and agencies of other countries)[a]	26.4
Total	100.0

SOURCE: *Waga-Kuni Sekiyu-Kaihatsu no Gengyo* [The Present Situation of Japan's Oil Development], Petroleum Producers Federation: Tokyo, 1974), p. 5. Original data from MITI.

[a]For a discussion of DD and GG oil see Chapter 4, section. 1.

as before World War II. Some Japanese suspected but evidently could not prove that diversions actually took place during the Arab oil embargo; many also remember the Allied blockade of Japan in World War II.

Blockade is but an extreme form of the disruption of sea-lanes. The geographical concentration of Japan's oil supply in the Middle East makes certain channels and choke points on the sea-lanes leading to its home islands of special interest to Japan. The Malacca Straits, Lombok, and the Bashi Channel, to mention a few, must be kept open, and the immediate land areas kept out of hostile hands. Yet even in this regard there is some uncertainty, albeit unrelated to the Arab embargo.

There were, of course, other adverse circumstances, notably the price hike ordered by OPEC members as a whole. In this connection, the value of Japan's crude oil and petroleum products import in 1972 was $4.46 billion.[6] The amount rose to $6.72 billion in 1973, registering a 50 percent increase in dollars. (See Table 8 for yen value.) In 1974, when the higher

TABLE 8

JAPAN'S IMPORT OF OIL AND PETROLEUM PRODUCTS
AND TOTAL IMPORT
1954–74

Calendar Year	Total Import (I)	Crude Oil	Petroleum Products	Subtotal Value (II)	Percentage Increase (preceding year = 100)	Ratio of II : I
		(in billion yen)				(percent)
1954	863.8	48.2	25.3	73.5	2.4	8.5
1955	889.7	53.5	30.3	83.8	14.0	9.4
1956	1,162.7	80.6	35.4	116.0	38.4	10.0
1957	1,542.1	116.5	64.5	181.0	56.0	11.7
1958	1,091.9	119.3	31.0	150.3	−17.0	13.8
1959	1,295.8	138.5	30.2	168.7	12.2	13.0
1960	1,616.8	167.4	48.7	216.1	28.1	13.4
1961	2,091.8	193.9	72.3	266.2	23.2	12.7
1962	2,029.1	223.4	77.9	301.3	13.2	14.8
1963	2,425.0	284.0	83.9	367.9	22.1	15.2
1964	2,857.7	334.4	91.1	425.5	15.7	14.9
1965	2,940.8	376.9	104.0	480.9	13.0	16.4
1966	3,428.2	432.1	98.8	530.9	10.4	15.5
1967	4,198.7	524.6	122.7	647.3	21.9	15.4
1968	4,675.4	606.7	149.4	756.1	16.8	16.2
1969	5,408.5	686.5	140.9	827.4	9.4	15.3
1970	6,797.2	804.8	198.0	1,002.8	21.2	14.8
1971	6,910.0	1,066.7	202.1	1,268.8	26.5	18.4
1972	7,229.0	1,209.7	166.0	1,375.7	8.4	19.0
1973	11,386.0	1,633.0	202.7	1,835.7	33.4	16.1
1974	19,130.0	5,503.9	656.5	6,160.4	235.6	32.2

SOURCE: The Miti White Paper, 1974. Also quoted in the *Seikiyu Kaihatsu Kankei Shiryo* [Data on Oil Development], PPF. For 1974, see *Monthly Statistics of Japan*; according to the Committee for Economic Development, Japan's petroleum import bill in 1974 came to $21.2 billion or 34.1 percent of total import. See *Consequences of High-Priced Energy* (New York, 1975).

price prevailed during the entire year, the total oil import bill—including both crude oil and petroleum products—was first expected to rise to $17.5 billion, but in the end came to $21.2 billion, or 34.1 percent of total imports and a 215 percent increase in dollars over the previous year's figure. Obviously, if international receipts should fail to rise sufficiently to cover the much higher oil bill and import of other goods including raw materials had to be cut back, exports might actually fall. Furthermore, exports could also fall as a result of the decline of external demand in other oil-consuming countries. Fortunately for Japan none of these contingencies materialized although the threat seemed real enough at the outset.

Actual Developments in 1973–75

What actually transpired in regard to Japan's oil import can be analyzed with respect to both (a) the immediate crisis and postcrisis months during November 1973 to early 1974 and (b) the postcrisis year of 1974–75. First of all, since the most serious threat posed by the oil embargo was a sharp, immediate curtailment of supply, it is interesting to note that Japan's crude oil import from all sources actually increased by 17.2 percent from 246 million kiloliters in FY 1972 to 288.5 million kiloliters in FY 1973. Since the Japanese fiscal year extends from April of the given year through March of the following year, the second half of FY 1973, or October 1973–March 1974 inclusive, encompassed the peak of the oil crisis in 1973 and the three months immediately thereafter. On a calendar year basis, the import figures in 1972 and 1973 were 249.2 million and 289.7 million kiloliters respectively.[7] This was an increase of 16.2 percent in one year. Furthermore, the aggregate amount of imported petroleum *and* products—including crude oil, LPG, and various distillates—rose by 15.2 percent, from 2,571,580 million kilocalories in FY 1972 to 2,962,180 million kilocalories in FY 1973.[8]

An original March 1973 MITI forecast of crude oil import in FY 1973 was 282.7 million kiloliters.[9] This estimate was adjusted downward to 259 million kiloliters in early December 1973 as a result of fears of a serious oil shortage due to the embargo. The March forecast for FY 1973 was made up by 130.6 million kiloliters for the first half of the fiscal year and 152.1 million kiloliters for the second half.[10] The December forecast, as the following calculations show, indicated a sharply lowered estimate for the second half. Actual crude oil import during the second six-month period turned out, however, to be 143.4 million kiloliters as compared with 145.5 million kiloliters in the first half of FY 1973. Thus not only was the import total during FY 1973 considerably higher than in FY 1972, but the real decline in the second half was a reduction of only 8.7 million kiloliters in comparison with the optimistic March estimate, and only 2.1 million kiloliters short of the actual import during the first six months of the fiscal year prior to the oil crisis. The revised December 1973 forecast was, therefore, far too pessimistic, especially for the first three months of 1974.

FORECAST OF CRUDE OIL IMPORT

(in million kiloliters)

	April 1973 to September 1973	October 1973 to March 1974	Total
March 1973 forecast	130.6	152.1	282.7
Actual	145.5	143.4	288.9
Actual less forecast	14.9	−8.7	6.2
December 1973 forecast	145.5[a]	113.5[b]	259.0

[a]Assumed to be known by the time of the December estimate.
[b]Obtained as a residual between 259.0 and 145.5.

Actually a postcrisis decline of crude oil import occurred only in the second half of the 1974 calendar year, well after the end of any embargo threat. As a result, the volume of crude import in the 1974 calendar year as a whole fell to 278.4 million kiloliters as compared with 289.7 million kiloliters for 1973.[11] This was a drop of nearly 4 percent. The decline then continued through most of 1975.

Further insight into the OAPEC threat of oil cutback and its aftermath can be seen by examining the monthly import data for several of Japan's major sources of oil import (Table 9) beginning from October 1973 when the threat was first made.

It takes about forty days for a tanker to make a round trip between the Persian Gulf and Japan. Since the embargo was first announced in mid-October, November 1973 would be the earliest month when its effect might become detectable in Japan's trade statistics. On the other hand, since the embargo was lifted at the very end of 1973, the resumption of large oil imports, if any, should be seen in Japanese trade returns in January and February of 1974, or even a little later if we allow for some delay. Since no further compulsory curtailment of supply took place afterwards, the import volume should not again fall on that score.

If we examine the data for Saudi Arabia and Kuwait given in Table 9, two different patterns are discernible. First, in the case of Saudi Arabia, Japan's crude oil import from that country fell during November, December, and January below the 6 million kiloliter monthly level registered in

October 1973. However, it quickly picked up in February 1974, and then for another eighteen months hovered between 6 and 7 million kiloliters each month, dipping slightly below the 6 million figure six times, mostly in 1975. The pattern in November 1973–February 1974 agreed with what one should expect as described in the preceding paragraph. However, in the case of Kuwait, Japan's crude oil import in November 1973 through

TABLE 9

Japan's Crude Oil Import
October 1973–August 1975

(in million kiloliters)

	Saudi Arabia		Kuwait		Iran		Indonesia	
	Monthly	Quarterly	Monthly	Quarterly	Monthly	Quarterly	Monthly	Quarterly
1973								
Oct.	6.1		1.6		8.8		3.8	
Nov.	5.4		2.4		7.5		3.5	
Dec.	5.8	17.3	2.0	6.0	7.8	24.1	3.9	11.2
1974								
Jan.	5.4		2.0		6.2		3.5	
Feb.	6.3		2.3		5.5		3.5	
Mar.	7.0	18.7	2.7	7.0	5.3	17.0	3.4	10.4
Apr.	7.1		2.1		6.2		3.6	
May	6.1		2.8		5.4		3.2	
June	6.1	19.3	3.2	8.1	5.1	16.7	3.2	10.0
July	7.1		3.0		6.4		3.5	
Aug.	6.8		1.8		5.7		3.4	
Sept.	5.5	19.4	1.6	6.4	5.6	17.7	2.6	9.5
Oct.	7.1		2.1		6.0		3.1	
Nov.	51.		2.0		6.5		B.T	
Dec.	6.2	18.4	2.3	6.4	7.0	19.5	2.9	8.9
Total	75.7[a]		27.9		70.9		38.9[a]	
1975								
Jan.	5.9		2.1		6.5		2.9	
Feb.	5.4		1.5		6.3		2.8	
Mar.	6.6	17.9	2.2	5.8	7.2	20.0	3.1	8.8
Apr.	6.0		2.1		5.6		2.8	
May	7.5		2.2		6.4		2.2	
June	5.7	19.2	2.0	6.3	4.3	16.3	2.1	7.1
July	6.9		2.1		5.3		2.1	
Aug.	5.9		1.6		5.3		2.5	

Source: *Monthly Statistics of Japan* (Tokyo), March and October, 1975.
[a]The monthly figures do not add up to the totals for the year due to rounding.

February 1974 was at a very steady level of 2 to 2.4 million kiloliters a month, all considerably *above* the 1.6 million kiloliter level of October 1973. Then, in four of the five succeeding months beginning in March 1974, import from Kuwait also rose—as was the case with Saudi Arabia—to above the usual level. However, during August 1974–August 1975, the monthly import again fell to around 2 million kiloliters or less.

The two other major suppliers of crude oil to Japan, *not* members of OAPEC, showed still more interesting patterns. In November and December 1973, Japan's monthly import from Iran fell to below 8 million kiloliters from a high of 8.8 million in October 1973. However, after having declined to below 6 (or around 5.5) million kiloliters a month in February and March, no subsequent large increase was registered after the removal of the Arab oil embargo. It was rarely above 6 million kiloliters except in the last quarter of 1974 and the first quarter of 1975. In the case of Indonesia, there was a drop from 3.8 million kiloliters in October 1973 to 3.5 million kiloliters in November. The amount then quickly went back to 3.9 million kiloliters in December and thereafter remained at a steady level of 3.2 to 3.6 million kiloliters through the entire first eight months of 1974. It then declined to around 3 million kiloliters for the rest of 1974 and the first quarter of 1975. A further drop again took place in the second quarter of 1975.

What these patterns seem to suggest are the following: First, Saudi Arabia registered a noticeable cutback of its crude oil supply to Japan during the crisis period in the last quarter of 1973. However, this curtailment was very promptly made up after the embargo was lifted. In the case of Kuwait, another Arab oil producer, there was no reduction at all even during the embargo. On the other hand, here too there was an increase immediately afterwards. It would seem, therefore, that the reduction ordered by OAPEC was not really uniformly applied. In fact, as the reader will recall, the OAPEC order of October 16, 1973, did provide certain loopholes in terms of the individual members' need to safeguard their special national interests and commitments. Moreover, the increase in Japan's import from these two Arab suppliers during several months after the embargo was probably linked to an effort to build up inventory. Some of this inventory-building may reflect successful Japanese efforts during the crisis period, when short supply was feared, to contract for long-term purchases at rather high prices in the then competitive scramble for oil. As for imports from Iran and Indonesia, the continuing lower level of import or decline extending beyond the crisis period implies that the threatened Arab oil cutback had very little to do with Japan's oil import from these two non-Arab sources.

The real decline of import came in the latter part of 1974, and it appeared to be more an outcome of the reduction of demand rather than supply, which should be attributed to the direct as well as indirect effects of the sharp rise of the oil price, and of the reduction of real national output in Japan *and* elsewhere in the world, the latter adversely affecting Japanese exports. Variations of Japanese import from different suppliers then reflected shifts of purchase to allow for quality, price, transport cost, and other differences.

When the oil cutback was first announced in October 1973, the posted price of Arabian Light was $5.119 per barrel. This rose to $5.176 on November 1, 1973, dipped slightly to $5.036 on December 1, but then leapfrogged to $11.651 on January 1, 1974. Other similar increases from October 16, 1973, to January 1, 1974, are illustrated below:

	October 1, 1973	October 16, 1973	January 1, 1974
	(posted price in dollars per barrel)		
Arabian Heavy	2.727	4.633	11.441
Iranian Light	2.995	5.341	11.875
Iranian Heavy	2.936	4.991	11.635
Sumatra Light (formula sales price)	4.750	–	10.800
Libyan	4.604	8.925 (Oct. 20)	15.768

SOURCE: *Data on Oil Development* (1974), p. 113. See also Table 1. The posted price of Arabian Light was lowered to $11.25 in November 1974 and was again hiked to $12.38 on October 1, 1975. See *Oil & Gas Journal* (Tulsa), November 25, 1974, p. 44, and *The Wall Street Journal*, November 3, 1975.

In the face of the increase in posted prices, the average price of Japan's oil import (CIF) went up accordingly, from $3.46 a barrel in October 1973 to $4.09 in November, $5.02 in December, $6.60 in January 1974, $9.90 in February, $10.51 in March, $10.94 in April, and $11.27 in May, staying at above $11.30 in the rest of the year. The spark was provided for a cost-push inflation.

It seems that some of the economic effects mentioned earlier in this study took place as expected. Coming on top of an inflation that had run on for some time, Japan's wholesale price index registered in the last quarter of 1973 an *annual rate* of increase of 39.7 percent—compared with 22.5 percent in the third quarter, 14.1 percent in the second quarter, 20.9 percent in the first quarter, and much lower rates of increase during 1972 (see Table 10). This price jump was followed by an even larger increase in the first quarter of 1974, when the wholesale price index rose at an *annual rate* of 72.5 percent. Following the rise of wholesale prices with a lag, the consumer price index took a leap in the first quarter of 1974 to an *annual rate* of 43.3 percent from that of 13.2 percent in the last quarter of 1973.

Simultaneously with these price increases, production and employment dropped. Japan's GNP at 1970 prices registered during the first quarter of 1974 a 14.1 percent decline at annual rate; industrial production in the

TABLE 10

SELECTED JAPANESE ECONOMIC INDICATORS DURING
AND AFTER THE 1973–74 OIL CRISIS

(Compounded annual rates of change)

(Preceding quarter = 100)

Indicator	1973		1974			
	3rd Quarter	4th Quarter	1st Quarter	2nd Quarter	3rd Quarter	4th Quarter
Wholesale price index	22.5	39.7	72.5	14.1	12.2	4.7
Consumer price index	12.9	13.2	43.3	21.4	14.6	18.1
GNP at 1970 prices[a]	0.4	Nil	−14.1	8.0	3.2	−1.8
Industrial production[a]	10.7	12.4	−8.0	−8.2	−13.5	−19.2
Manufacturing employment[a]	0.8	1.6	−2.4	1.6	−2.4	−4.8
Money supply[a]	18.0	3.0	18.8	24.1	−3.2	8.2
Imports (CIF)	44.5	72.5	218.9	57.3	3.6	−3.7
Exports (FOB)	15.7	71.7	93.4	107.1	31.0	34.9
(in billions of yen)						
Imports[b,c]	2,697.7	3,091.5	4,131.2	4,626.5	4,667.7	4,623.9
Exports[b,c]	2,449.5	2,803.8	3,306.6	3,966.9	4,243.8	4,573.4
Balance[c] (export less import)	−248.2	−287.7	−824.6	−659.6	−423.9	−50.5
(in billions of dollars)						
International reserves	14.80	12.25	12.43	13.43	13.17	13.52

SOURCE: Federal Reserve Bank, San Francisco (FRB), *Pacific Basin Economic Indicators* (San Francisco, May 1975), pp. 29–34. For the unadjusted trade date of 1974 and 1975 in dollars, see Bureau of Statistics, *Monthly Statistics of Japan,* October 1975 and earlier issues quoted in the text.

[a]Seasonally adjusted in IMF, *International Financial Statistics.*

[b]Seasonally adjusted by FRB, San Francisco.

[c]The unadjusted original data would show an export *surplus* of $662 million in the last quarter of 1975. If imports and exports are both valued at FOB, thus segregating transportation of import to a different category, the trade figures and the corresponding current account balances in 1972–75 would be (in millions of dollars):

Year	Import	Export	Trade Balance	Current Account Balance
1972	19,061	28,032	+8,971	+6,624
1973	32,576	36,264	+3,688	−136
1974	53,044	54,480	+1,436	−4,693
1975	49,714	54,822	+5,108	−680

SOURCE: Bank of Japan and IMF (1975 figures are preliminary).

same period fell at an annual rate of 8 percent; manufacturing employment, at 2.4 percent. The cost-push inflation was sustained by an increased rate of expansion of money supply during the first half of 1974 which was, however, essentially the continuation or—better still—resumption of expansion that had been carried on for already more than a year. At annual rates, the money supply expanded at 18.8 percent in the first quarter and 24.1 percent in the second quarter of 1974. (The annual rates of increase in 1973 were 24.5 percent in the first quarter, but only 3.0 percent in the fourth quarter.)

Although Japan was taken off the embargo list, the high price of oil continued. Double-digit inflation in consumer prices also continued through 1974, albeit at a far more moderate rate than in the first quarter. Only in 1975 did the annual rate of increase fall to below 10 percent, with the exception of the third quarter when the increase was at an annual rate of 17 percent. As a countermeasure to the rising inflation, money supply was tightened and even declined during the third quarter of 1974. Throughout 1975 money supply increased at an annual rate of from 12.2 percent in the first quarter to 8.3 percent in the third quarter. The rise of wholesale prices finally dropped to an annual rate of 4.7 percent during the last quarter of 1974, and actually declined during the first two quarters of 1975. In the last quarter of 1975, wholesale prices were again rising at an annual rate of 4.7 percent, moderate by recent standards.

Industrial production declined at an increasing rate during 1974; the annual rate of GNP growth at 1970 prices, which had again become positive in the second quarter of 1974, slowed during the second and third quarters, finally turning to a negative 1.8 percent in the last quarter. Both industrial production and GNP did not turn around until the second quarter of 1975. Manufacturing employment, after a brief recovery in the second quarter, resumed its decline during the third and fourth quarters of 1974. The decline, at a higher rate, continued through the first half of 1975, slowing down only in the third quarter.

The net effect of the contraction contributed to an improved position on Japan's commodity trade account for a time during the last part of 1974, reversing the earlier deficit. An export surplus, at the level of $662 million, again emerged in the fourth quarter of 1974, and only a small decrease in international reserves took place as compared with the third or pre-oil-crisis quarter of 1973. Beginning from October 1974, with the exception of the month of December 1974, the current account balance in Japan's international balance of payments (seasonally adjusted) also turned favorable and remained so through March 1975. The same trend prevailed also in the case of the overall balance. There was also a definite improvement in both the current and the overall balance for all 1975 compared with a year earlier. In a report of the Economic Planning Agency to the Council of Economic Ministers, dated April 16, 1975, foreign currency reserves were given as $14,152 million at the end of March 1975, having regained the level of October 1973 when the oil crisis began, although it again fell to $12,815 million at the end of December 1975. On the other hand, for the entire year of 1974, import of crude oil and petroleum products cost Japan $21,162 million in contrast to $6,726 million a year earlier. The trade deficit in 1974, *valuing imports at CIF*, equalled $6,575 million in comparison with a $1,384 million deficit in 1973. Furthermore, the $662 million trade surplus of the fourth quarter in 1974 turned into a deficit of $1,241 million in the first quarter of 1975 and a smaller deficit of $623 million in the second quarter.[12] For the entire year of 1975, Japan had a trade deficit between exports (FOB) and imports (CIF) of $2,110 million. The national disaster in international payments envisaged at the time of the oil crisis was successfully weathered for the time being, but the success was by no means irreversible or unqualified. 1990556

Improvements in Japan's payments position was an outcome of domestic policy and external developments, and it is interesting to note Japan's adoption of a policy of monetary restraint, especially in the second half of 1974, and the apparent emphasis on price stability and the external balance and the concomitant decrease in output and employment and increase in business failures. However, the 1975 worldwide recession and declining demand for oil import also exacted their toll on the solidarity of the oil cartel.[13] An OPEC meeting in September 1975 again agreed to raise the posted price by 10 percent at the insistence of Iran and others in spite of a reluctant Saudi Arabia and Iran's original proposal for a 15 percent increase. However, several countries either failed to carry out the full 10 percent increase or unilaterally made price cuts soon after publicly announced increases. For instance, in October 1975 Indonesia changed its posted prices for various grades of crude to a range of $11.50 to $13.00 a barrel, which was less than the 10 percent increase ordered by OPEC the same month. Algeria raised its posted price from $11.75 to $12.75, also by less than 10 percent. In Saudi Arabia, although Arab Light was posted at $12.38 a barrel in October, which represented a 10 percent increase, the new posted price of Arab Berri was only $12.77, or 6.8 percent higher than before. Kuwait raised the posted price of its heavy crude first to $11.40 during October, but cut the price back only a month later to $11.30. Even Iran was forced to cut back its posted price in February 1976 from $11.50 to $11.40. By varying premiums for differences in grade, quality, and transportation cost, other OPEC members were equally ready to undercut one another as a result of declining revenue from reduced sales. Under-the-counter price shading was allegedly practiced by Indonesia and Iraq long before the September 1975 meeting of OPEC, while Nigeria and Libya were accused of premium manipulation. However, no one was as yet willing to dip substantially below the much higher price established under OAPEC's original embargo threat.

All in all it would appear that a succession of events with opposing effects on Japan's oil bill and balance of international payments have occurred since the winter of 1973. The high oil price, the domestic recession, and the policy of monetary restraint to curb inflation reduced the demand for oil import and, therefore, overall import payments to below the level they might otherwise have reached. The worldwide recession then had an adverse effect on Japan's exports, which both further reduced demand for oil and raw material import and lowered export income. At the same time, the inflated prices of Japanese exports have tended to increase the country's export earnings where competitive advantage has not been seriously impaired, partly because of even greater inflation in other countries or where Japanese relative advantage has been maintained through extra efforts in export promotion. This has more than compensated the otherwise unfavorable effect on exports. Finally, there has been some cracking of the solid price front of the oil cartel. Outside the merchandise trade account, there was a cutback of Japan's capital outflow both because of recession in potential host countries and as a result of the stricter control and the decline of liquidity on the part of potential Japanese investors. In addition, there has been an inflow of some oil money for investment in Japan. The timing and disparate strengths of these forces have produced a checkered pattern so that a short- to medium-term evaluation of the net effect of the high oil price on Japan's payments position would yield results varying with the time the analysis is made while the net long-term effect is by no means clear. However, Japan has certainly *more than survived* the oil crisis.

As far as the quantity of oil supply is concerned, no shortage has actually occurred either during the 1973 crisis or at any time thereafter in the subsequent two years. On the other hand, Japan's potential vulnerability to a supply cutoff will remain and there is no visible end to the need for a reduction of demand through conservation, changes in the industrial structure, technology, and rate of economic growth, and development of alternative sources of energy and oil supply. Of these policy goals, development of more secure alternative sources of oil supply is the most important as long as dependence on oil can be reduced only in a very limited fashion within, say, the next decade.

3

INTERACTION WITH
MOSCOW AND PEKING

Even though the Arab embargo did not really cut off Japan's oil supply, it highlighted the country's vulnerability and gave considerable impetus to the long-advocated diversification program. Besides, it is both economically and strategically desirable to develop sources of supply close to the Japanese home market. If such a development could contribute to diversification, it would be doubly desirable. In this respect the Soviet Union presents itself to Japan as an economically promising source of many producer goods. Not only does it possess potentially very large quantities of natural resources in general, but the prospects appear especially favorable in the case of energy products. Some Japanese students of the Soviet Union believe that the latest Soviet economic plan to develop Western Siberia as a major base of production of industrial raw materials and fuel will make available an increasing exportable surplus some of which can be sold to Japan.[1] As a matter of fact, ever since the initiation of periodic bilateral discussions on the Joint Japan–Soviet Economic Committee in 1966, the prospects of developing Siberia with Japanese financial and technical help in return for promised future supplies of, for example, timber, coking coal, natural gas, crude oil, and refined petroleum products, have been continually emphasized by the Soviet Union. Against a background of potential shortages, this prospect of more Soviet oil in the not-too-distant future appears particularly hopeful. Since the Soviet Union supplied only 0.2 percent of Japan's crude oil import in 1972,[2] a sizable increase would also be most desirable from the point of view of the intensified diversification drive.

Parallel to these possibilities presented by the Soviet Union, the People's Republic of China (PRC) offers no less rosy prospects from the perspective of Japan. The apparent promise of the China market—both as a traditional source of supply of minerals and other crude materials and, in the eyes of some exporters, as a potentially insatiable market for Japanese capital goods and other manufactures—has always attracted the attention of Japanese businessmen. The Japanese appetite was whetted in the early 1970s by reports of record Chinese crude oil production. For example, China's crude oil production rose from 20 million tons in 1970 to around 50 million tons in 1972. Similarly, Chinese natural gas production reportedly rose from a mere 10 billion cubic meters in 1967 to 45 billion cubic meters in 1973. Two of the more recently discovered large producing fields, Ta-kang and Sheng-li, are located on opposite sides of the Pohai Gulf where off-shore deposits are reported to be most promising.[3]

Politically, the radical change in U.S. policy toward China in 1971 sharply enhanced the urgency of Japan's need for other options than total dependence for its national security upon the United States. The emergence of this new requirement, which had in fact been evolving for quite some time, was superimposed upon Japan's growing desire for more assertiveness and recognition by other nations as a rising power in world affairs, independent of its erstwhile American mentor. From Japan's point of view, a principal if not the only viable choice to safeguard Japan's national security, short of accelerated large scale rearmament—which would probably meet with external suspicion in Asia and overwhelming internal opposition in the prevailing conditions of the early 1970s—was to try to bring about a balance of interests among the big powers so that none would bear ill will toward Japan. To achieve such a state of equilibrium would require, *inter alia*, that Japan's continued independence and economic well-being be very useful to each of the big powers. In particular, therefore, Japan would have to expand its role as a supplier of goods to both the Soviet Union and the PRC, its importance to the U.S. economy being already firmly established. This is true especially in the Soviet case, because as of 1971, when the new U.S.-China policy was instituted, Japan's total trade with the Soviet Union was not more than 3 percent of total Soviet foreign trade and only 2.2 percent of Japan's total foreign trade.[4]

In the case of the PRC, Japan was already the largest supplier of China's imports. Total Sino-Japanese trade[5] in 1972 was 19 percent of China's external trade, albeit only 2.1 percent of Japan's total foreign trade.[6]

Over and above Japan's long-standing interest in promoting exports as a whole, additional new reasons call for a special Japanese effort to increase trade with China. Of particular concern within the context of this study, Japan would like to use the expansion of Sino-Japanese trade and other bilateral economic ties as leverage in negotiations with the Russians, as well as with the Chinese themselves. Japan, too, it appears, would wish to use the ploy of Nixon and Kissinger in 1971–72 in exploiting the mutual hostility and suspicion between Moscow and Peking to its own advantage.[7] Japan also had a strong incentive to strike a stance of independence in its own China policy that would overshadow the American initiative. An increasing volume of trade would be a good bargaining counter in negotiating with Peking. The possibility of approaching China on oil became an attractive reality in view of recent reports on the increase in Chinese oil production.

Negotiation with the Soviet Union
for Oil and Other Energy Products

Japanese direct purchases of crude oil from the Soviet Union were at a low level in the early 1970s. According to the Japan Petroleum Federation, import of crude from the Soviet Union decreased steadily from 577,000 metric tons in 1970 to 461,000 tons in 1971 and 392,000 tons in 1972.[8]

These imports came from Sakhalin and were augmented by Middle Eastern oil through swapping, that is, oil delivered (as it were) "for the account of the Soviet Union" by non-Soviet producers. During this period, the latter imports first rose from 1.354 million tons in 1970 to 1.991 million tons in 1971. The amount then fell to 1 million tons in 1972. Thus, on the eve of Japan's stepped-up oil negotiations with the Soviet Union, there had been a decline in Japan's total (i.e., both direct and indirect) import of Soviet oil, from 2.452 millions tons in 1971 to 1.392 million tons in 1972.

Japan's direct import of Soviet crude oil increased in 1973. This develop-ment appeared to be partly a Soviet response to the initiation by Japan of negotiations with the PRC for the importation of Chinese oil. There is evidence of considerable interaction between concurrent Japanese-Soviet and Japanese-PRC negotiations. The Japanese-Soviet oil negotiations were focused on Tyumen and Sakhalin. However, these two projects were only a part of a much broader program of bilateral economic cooperation, discussion of which by the two countries began a decade earlier.

The genesis of the Japan-Soviet Economic Committee, which convened its first meeting in 1966, can be traced back to Mikoyan's 1961 visit to Japan. At the second meeting of the committee in June 1967, the possibil-ity of Japanese cooperation in the development of Tyumen oil was brought up for discussion by the Soviet Union. Two other projects, port construc-tion at Wrangel and the development of Soviet Far Eastern timber resources, were also on the agenda. Other projects in the energy field came up during subsequent meetings of the committee, including the Yakutsk Natural Gas Project at the third meeting in January-February 1968, the Sakhalin Oil and Gas Project at the fifth meeting in February 1972, and the South Yakutsk Coking Coal Project, also discussed in 1972.[9]

Although the Tyumen project was first brought up for discussion by the Soviet Union in June 1967, little real progress was made during the next few years. It was not until the beginning of 1972, when coking coal from South Yakutsk and oil and gas from Sakhalin were added to the list of tantalizing Soviet proposals for Japanese consideration, that discussion on the Tyumen project took on a new light. Although one can only speculate on Soviet motives, a plausible political explanation of the particular timing was the Soviet Union's desire to develop better relations with Japan in order to discourage closer relations between Tokyo and Peking, perhaps even to erode U.S.-Japanese relations. Following soon after the initiation by the United States of its rapprochement with Peking, such a Soviet move toward Japan would seem to be eminently reasonable. At any rate, Soviet expressions of its readiness to proceed in earnest with the issue of economic cooperation was greeted by considerable enthusiasm on the Japanese side, given the special circumstances of Japan's oil supply we have already described.

Early in February 1973, the Soviet Union requested Japan to clarify its position on possible bilateral cooperation in the further development of Tyumen. In response, and following an official declaration of policy in the Diet, Hiroki Imasato, president of the Overseas Oil Development Company and chief Japanese delegate to the Oil Subcommittee of the Japan-Soviet Economic Committee, publicly expressed the hope that earnest negotiations would begin in Moscow in April 1973.[10]

Then, early in March, Prime Minister Tanaka told Brezhnev in a personal letter that the Japanese government was "greatly interested in the development of Siberia" and that it would deal with this matter "with a forward looking attitude," pending the completion of technical talks between the parties concerned. This Japanese offer was tied to the "continuation of negotiations on a Japan-Soviet peace treaty during 1973." Tanaka took this initiative after Foreign Minister Ohira had failed to engage Brezhnev in discussions on the peace treaty during Ohira's 1972 visit to the Soviet Union.[11] A linkage between economics and politics was very deliberately established.

On March 12, 1973, Nakasone, then head of MITI, again took the initiative by inviting Troyanovsky, the Soviet ambassador, to a meeting at which the former clearly indicated Japan's forthcoming attitude toward granting bank loans for the Tyumen project.[12] This step was a part of the implementation of Tanaka's policy expressed in his letter to Brezhnev.

This series of steps taken in rapid succession by high Japanese officials represented a concerted effort to explore the possibilities of future relations with the Soviet Union. Having "normalized" its relations with Peking in the fall of 1972, Japan was now ready to pursue its Soviet probe. The Japanese grand design seemed to be aimed at the construction of a network of balanced economic relations with Peking and Moscow in which the United States would be a partner on the Japanese side. Such a structure would, it was hoped, produce an equilibrium safeguarding Japan's national interest.

Nakasone's proposal to the Soviet Union was to proceed with simultaneous negotiations on Tyumen oil, Yakutsk natural gas, and Sakhalin oil and gas. At this juncture, the three projects were reported by Japanese sources to require a total credit of about $4 billion from the Export–Import Bank of Japan: about $1 billion for the Tyumen project, which would include the construction of a 4,300-kilometer pipeline to the port of Nakhodka; $2 to 3 billion for the Yakutsk Natural Gas Project, which would also include pipeline construction and the manufacture of LNG at Nakhodka for further shipment to both Japan and the United States; about $230 million for the Sakhalin project.[13] One of the contested issues in the negotiation concerned the rate of interest: the Japanese were asking for 6.5 percent while the Soviet side proposed a rate of 6 percent in addition to a four-year deferment of repayment.[14]

As will be seen later, discussions on the initiation of oil import from China were being carried on by Japan at the same time. These activities were being publicized widely—perhaps deliberately—and the Russians were undoubtedly fully aware of them. At any rate, Idemitsu Kosan, which had always been an importer of Soviet oil, succeeded in contracting for the direct import of 1 million tons of Tyumen oil in 1973. This was an increase of more than 150 percent in such direct import over the 1972 level.[15] Previously the Soviet Union had curtailed oil shipments from Tyumen to Nakhodka perhaps partly as a means to exert pressure on Japan in ongoing Soviet-Japanese negotiations. After all, the Soviet Union was urging Japan to help construct a pipeline as a more economical and reliable mode of transporting oil to Nakhodka. Japanese tankers would provide onward shipment from Nakhodka to the Tokuyama refinery of Idemitsu. According to the speculation of *Nihon Keizai*, a principal Soviet motive in agreeing to this sale was to discourage Japanese negotiations with the PRC.[16]

In June 1973, Idemitsu followed up its earlier success by raising its target of future import of Soviet oil. This was now set at 3 million tons for 1974, which would be a little over 1 percent of Japan's total import. Actual import since April 1973 was then running at 150,000 tons a month, or an annual rate of 1.8 million tons.[17] Another report indicated that Idemitsu hoped to import in 1974 at least 2 million kiloliters of Tyumen oil and that the amount might even reach 3 million kiloliters.[18]

By the early summer of 1973, however, the initially favorable indications began to change for the worse. The Japanese, in trying to get discussion on Tyumen started, were now encountering a series of Soviet delays. Originally scheduled to open on May 21, the sixth meeting of the Japan-

Soviet Economic Committee was first postponed to June, then to July, and again to August.[19] Japanese participants either believed, or were told by the Russians, that the Soviet Union needed additional time to complete more detailed surveys of available resources at Tyumen. They also thought that the Soviet team's homework included considerations of security of the pipeline in the light of prevailing hostile Sino-Soviet relations. Furthermore, the Japanese government was said to be expecting a higher Soviet loan request than the figures previously envisaged. According to one source, the new figures might be $2 billion for the Tyumen Oil Project, $600 million for the South Yakutsk Coking Coal Project, $3 billion for the Yakutsk Natural Gas Project—of which Japan's share would be $2 billion, the rest to come from U.S. financing—and $1 billion for the Sakhalin project. These individual items would total $5.6 billion as the possible total Soviet loan request to Japan.[20]

During discussions between the two sides in August 1973, the substance of the Soviet proposal finally became public. One major revision of the Soviet plan was a marked reduction of the potential supply of Tyumen oil to Japan. The Japanese side claimed that it had been originally given to understand that the output of crude oil at Tyumen probably could reach 125 million tons by 1975 and 230–250 million tons by 1980, and that Soviet export from Tyumen to Japan would also rise from 25 to 40 million tons a year along with expanding total production.[21] Furthermore, export to Japan at the 40-million-ton level would continue for twenty years. The revised Soviet proposal, however, indicated that Japan could expect to receive only 25 million tons a year from Tyumen even with production at its peak.

In addition, the Soviet Union reportedly was thinking of a change in the method of oil transportation, shifting from pipeline to tank cars.[22] However, the exact nature of this aspect of the Soviet plan did not become public knowledge until another few months later.

Initially, Japanese misgivings about changes in Soviet plans were focused on the threatened reduction of supply from Tyumen to below 40 million tons a year. Speculating on the underlying reasons of the proposed reduction, one explanation offered by Japanese analysts was that Soviet oil reserves and the projected supply from Tyumen may have turned out to be smaller than originally suspected, especially in view of the increased demand for Soviet oil from Eastern Europe, not to mention the USSR's own domestic demand.[23]

From the point of view of Japanese business interests contemplating investment in Siberia, a reduction of the supply from Tyumen would make the large Japanese capital commitment less attractive, especially if the magnitude of the loan involved was to be raised even as the promised

supply would be reduced. Apart from the official Japanese position publicly maintained by Prime Minister Tanaka, which continued to insist on
Soviet assurance of the original 40-million-ton target, a number of alternative suggestions were apparently discussed on the Japanese side. One of
these was to proceed with a Soviet proposal of providing Japanese financial
and technical aid to the construction of an oil refinery at Nakhodka in return
for Soviet supply of an additional 6 million tons of crude oil from Tyumen
every year. However, this Soviet proposal was not looked upon with favor
by some Japanese who had doubts about the total volume of crude oil that
would be available to meet Soviet domestic demand in the Far East, as
well as the demand for export to Japan. Still another idea discussed at this
time was to redouble Japanese efforts to increase oil supply from other
regions, including the construction of a large storage facility at Lombok in
Indonesia that might be jointly financed by Japan, the United States, and
Western European countries. Another idea was to increase Japan's financial participation in the development of North Sea deposits.[24]

From the Japanese public as well as private business point of view, the
increased Soviet loan request made financial participation by U.S. sources
even more desirable than before. An American share in financing would
have the further advantage of partly neutralizing the Soviet Union's leverage against Japan. Yet another political consideration on the Japanese side
was the desire to placate Peking. Apparently Chou En-lai had emphasized
in a discussion with Kogoro Uemura, chairman of the Keidanren, that
Japan's aid in the development of Tyumen oil should be undertaken with
U.S. participation rather than as a bilateral Soviet-Japanese enterprise. It
was against this background that Imasato of the Oil Subcommittee of the
Japan-Soviet Economic Committee undertook to discuss with Gulf Oil the
possibility of the latter's participation in the Tyumen project.[25]

Popular discussion in Japan in the late summer of 1973 did not evince
any anticipation of the impending oil crisis that was soon to be triggered
by the Yom Kippur war and the ensuing Arab oil embargo. The Soviet
Union may have reached a more pessimistic estimate of its own oil situation. With the benefit of hindsight, however, one wonders whether the
Soviet Union had not anticipated the Arab attack on Israel and its potential impact on the supply of oil to major oil-importing countries. Certainly
the Soviet Union was in the thick of creating the necessary conditions for
an Egyptian attack and encouraging the Arabs' use of their oil weapons.
The harder, revised Soviet position in discussions with Japan in August
1973 would not be inconsistent with Soviet anticipation of the war as a
factor in Soviet planning. The fact that, following the onset of the oil
crisis, the Soviet Union was promptly ready to pressure Japan toward
greater accommodation to Soviet wishes was made amply clear to Uemura

and Nagano in early 1974. Upon their return from discussions in Moscow in March 1974, the two Japanese business leaders disclosed to the press that the Soviet Union had proposed a different mode of transportation of oil from Tyumen. The Soviet plan now called for the transportation of oil by pipeline from Tyumen to Ust Kut (or Lena City) north of Lake Baikal, from which point it would be carried by tank car to Komsomolsk-na-Amure near the Soviet Pacific Coast (to the north of Khabarovsk and northeast of the Manchurian border) *on a new railroad.*[26] From Komsomolsk the oil would again be piped to the Pacific Coast port of Sovetskaya Gavani.

This proposal came at a time when Japan was still engaged in trying to inveigle the United States into participation in the Tyumen project in spite of the unavailability of credit from the U.S. Export–Import Bank (EX-IM Bank) as a result of adverse developments in the United States Congress on the new trade bill. Injection of the issue of free emigration of Soviet Jews had raised a new obstacle to the granting of MFN (most-favored-nation) status and credit to the USSR. Nonavailability of U.S. EX-IM Bank credit and guarantees had already stymied the earlier proposal of U.S. participation in the joint financing of the Yakutsk Natural Gas Project. Since the size of the proposed U.S. participation through Gulf Oil would be not more than two-tenths of the total amount involved, the Japanese were still hopeful that discussion on the Tyumen project could continue in spite of the U.S. Congress.

Nevertheless, in Japanese thinking, the new Soviet proposal of resorting to oil shipment by rail greatly magnified the political aspect. From Japan's point of view, while oil carried by pipeline could benefit the logistic position of the Soviet military in the Far East, including the Soviet Pacific fleet, a railway would be useful in transporting larger quantities of many other supplies as well as troops. Such an increase in transport capability would alter the overall strategic balance in the Far East between the Soviet Union and China and in addition might affect the naval balance in the Western Pacific between the Soviet Union and the United States. Japan was also most sensitive to potential objections Peking might raise.

Against this rising ferment, the Soviet ambassador to Japan attempted to introduce a little calm. Speaking at a press luncheon sponsored by the Nippon Kisha Club, Ambassador Troyanovsky stressed that construction of the new Baikal-Amur railway was a Soviet project begun some time ago and that it would be completed in about five years. He also stated that the Soviet loan request to Japan for the development of Tyumen oil was not a request for the joint construction of the new railway although loan proceeds might be used for purchasing material and equipment needed in the railway construction. He denied the effect of the new railway on international power relations, arguing instead that the most effective way for Japan to guarantee its national security was to enhance economic relations with the Soviet Union. Furthermore, stated Troyanovsky, the Soviet Union had only offered to export to Japan crude oil from Tyumen within the range of 25 to 40 million tons a year, and it had never promised to export 40 million tons for sure. Since demand for oil in the Soviet Union and Eastern Europe had expanded, Soviet supply of more than 25 million tons a year to Japan would be impossible. Nevertheless, the amount conceivably could be greater depending upon future developments.[27]

Some Japanese businessmen might have been willing to consider the Soviet railway proposal in spite of its military implications. At any rate they seemed to maintain the public position that these noneconomic factors were outside their province. Rather they were the concern of the Foreign Ministry.[28] Regardless of the ambiguity about the decision-making process in the formulation of Japan's foreign policy, which this attitude implied, the Tyumen project was shelved—for the time being if not indefinitely.

Although the Tyumen negotiations were held up, other projects to promote Soviet-Japanese economic cooperation forged ahead. In particular, the Yakutsk Natural Gas Project, the Coking Coal Project, and the Sakhalin Oil and Gas Project all came to varying degrees of fruition in April 1974. In addition, by the end of 1974 Idemitsu was again planning to import oil from British Petroleum through a swap arrangement with the Soviet Union.[29]

The Yakutsk Natural Gas Project was first proposed in early 1968 at the third meeting of the Japan-Soviet Economic Committee. The project envisaged an initial bank credit of $3.5 to $5 billion to be jointly supplied by the United States and Japan for the construction of a 3,500-kilometer pipeline from Yakutsk to the Pacific coast, as well as that of an LNG plant at or close to Nakhodka or, as an alternative, near the source of the natural gas supply. Beginning in 1979, according to the proposal, there would be an annual export of 10 to 15 billion cubic meters of natural gas to Japan, and an equal amount would be exported to the United States. On the Japanese side, discussion was headed by Anzai, head of Tokyo Gas; on the U.S. side, El Paso Natural Gas and Texas Eastern were reported to be the interested parties. The plan for joint financing by the United States and Japan was intended to follow the pattern set in the Murmansk Plan for bringing natural gas to the U.S. Atlantic coast. Active discussion of the extension of credit from the two countries took place in the winter of 1973–74. The initial difference between the Japanese and Soviet sides on the rate of interest was again one of 6 percent (as proposed by the Soviet Union) versus 6.5 percent (as proposed by the Japanese Export-Import Bank).

In spite of difficulties in securing U.S. participation for lack of EX-IM Bank facilities, by the end of April 1974 Soviet-Japanese negotiations finally resulted in the announcement of a master loan contract. Under this master plan, Japan would extend $1.05 billion of credit to the Soviet Union, including $450 million for coking coal development in South Yakutsk, $100 million for the Yakutsk Natural Gas Development Project, and $500 million for the second phase of the Far Eastern Forest Development Plan. The interest rate was settled at 6⅝ percent plus a 0.25 percent commitment fee. In addition, there was to be a 10 percent down payment on equipment purchased under the contract. The repayment schedule was as follows: completion of repayment in eight years, beginning in 1980, for the coal project; five-year repayment, with a grace period of three years, for the natural gas project. This master contract would become valid if agreement on the individual projects could be reached within six months.[30]

A general agreement was signed by Soviet and Japanese representatives on April 26, 1974, at the Keidanren Hall. The desirability of trilateral cooperation that would include the United States was again reaffirmed on this occasion. A further memorandum was signed four days later by the Soviet Union and Japan, the latter represented by Hisao Makita,[31] president of Japan Steel Tube and chairman of the Coal Subcommittee within the framework of the Japan-Soviet Economic Committee. It should be noted that the timing of these agreements followed immediately the impasse reached in the Tyumen negotiations. Although the Yakutsk Natural Gas Project was also shelved later[32] because of the unavailability of joint U.S. financing, the conclusion of the preliminary agreement in April and the parallel pursuit of other projects, notwithstanding the obstacles encountered over Tyumen oil and Yakutsk natural gas, demonstrated the desire of Japan and the Soviet Union not to let the entire program drop.

Also at the end of April 1974, the Japan–Soviet Economic Committee announced the signing of a memorandum by Hiroki Imasato, chairman of Japan's Special Committee on Sakhalin Continental Shelf Prospecting, and Sushkov of the Soviet Foreign Trade Ministry. The provisions of the memorandum included an initial Japanese credit of $100 million to cover the acquisition of equipment and facilities under a "lease formula" for the first five years, and the promise of a second credit installment of $100 million for another five years if results during the first phase would warrant the extension. In return for this credit Japan would be able to purchase oil produced from the project amounting to 50 percent of the total output, as well as an as yet undetermined amount of natural gas. The Soviet Union also promised to charge Japan a reduced price for the prospective oil and gas, reported to be 8.4 percent below the prevailing international market price at the time of future purchase.[33]

Furthermore, the agreement of April 27 stated that the Soviet Union would not object to participation on the Japanese side by U.S. enterprises as Japan's partners. This apparently was a device for the benefit of Japan in order to keep open the option of future U.S. participation without holding up the project because of the prevailing obstacles to U.S.–Soviet cooperation as a result of the snarled trade bill in the Congress. Under the agreement Japan promised to provide a certain sum to finance Soviet purchases of the computer systems needed for the prospecting operations. This device apparently bypassed the problem of possible COCOM (Coordinating Committee) control that might apply if U.S. computer systems subject to American export control were supplied to the Soviet Union by Japan unilaterally without prior U.S. agreement. The agreed arrangement presumably would absolve Japan of any responsibility if the Soviet Union managed on its own to acquire the needed computer systems. The protocol for the Sakhalin project was signed in Moscow on December 10, 1974; signing of the formal agreement took place again at the Keidanren Hall in Tokyo on January 28, 1975.

Under the December 1974 agreement on Sakhalin oil development, Japan would supply an initial credit of $52.5 million, including $30 million to finance on-the-spot expenditures, out of the total credit of $100 million set for the first five-year period. According to Japanese sources, the expected increased production in Sakhalin might reach 100 million tons of crude over ten years.[34] If this were realized, Sakhalin's annual rate of production would average 10 million tons. At the level of 50 percent of this total, export to Japan would amount to 5 million tons. This would approximately triple the volume of Japan's import from the Soviet Union (not counting the swap). While the Sakhalin prospects are considerably below the reduced—not to say the original—amount promised by the Tyumen project, it is a partial compensation for both Japan and the Soviet Union in the face of the stalled Tyumen negotiations. Whether this project can finally come to fruition, however, remains to be seen.

According to one Japanese estimate, the previously mentioned Soviet proposal to build a refinery at Nakhodka with Japanese aid envisaged an output capacity of 12 million tons a year. If this is added to the annual capacity of two small refineries at Khabarovsk, estimated at 2.5 million tons, the total would amount to 14.5 million tons. Against this potential supply the same Japanese estimate puts demand of the Soviet Pacific fleet at 8.5 million tons and that for other purposes in the Soviet Far East—usually met by production from Sakhalin—at 2.5 million tons. Thus, there would be an exportable surplus of approximately 3 million tons a year.[35]

A slightly later Japanese estimate raised the projected Soviet Far Eastern demand for petroleum products in 1980 from an original estimate of 7.5 million tons a year to 15 million tons. This increase was attributed to an expected expansion of Soviet military demand. If the original refinery capacity used in processing Sakhalin oil can provide no more than 2.5 million tons, the difference of 12 million tons would absorb the total capacity of the projected Nakhodka refinery and there would be no surplus of petroleum products available for export to Japan.[36] Some Japanese students of Soviet affairs speculate that even if the supply of Tyumen oil for the Soviet Far East *and* export to Japan should amount to 40 million tons, a major portion of this supply would be absorbed by increased Soviet domestic demand in the Far East, as well as rising Soviet military demand. Thus there may be very little left for export to Japan beyond the 25-million-ton level in the revised Soviet estimate. Hence, it behooves Japan to concentrate on the development of Sakhalin for the time being. The Sakhalin project would be less costly and would require no *direct* Japanese contribution to the construction of the Baikal-Amur railway. From the Soviet point of view, the Sakhalin project would constitute a first instance of Soviet-Japanese cooperation in offshore exploration on the continental shelf.[37] Thus both countries may find sufficient reason to proceed with the Sakhalin plan, that is, if new objections on national security grounds are not raised by the Soviet side.

On the whole, however, two years of intensive Soviet-Japanese negotiations had failed to yield a more glittering prize for either side. Japan had not succeeded in establishing a major new source of oil from the Soviet Union through investment participation.[38] The Sakhalin project marks only a small and very tentative beginning insofar as Japanese participation through direct investment in oil exploration is concerned. By the end of 1974, Japan was again reverting to the idea of buying oil directly from the Soviet Union, offering perhaps bank loans that are not tied to specific projects of investment in the Soviet Union by Japan. The Soviet Union, on the other hand, also failed to tie Japan more closely to its orbit by creating economic relations that it would be too painful for Japan later to break.

Oil Negotiations between Japan and the PRC

Japan's crude oil import from the PRC, unlike its import of Soviet oil, took place for the first time in 1973. This became possible as a result of the sharp increase in Chinese crude oil production beginning in the second half of the 1960s. Chinese crude production in 1960 has been estimated at about 5 million tons.[39] By 1965, on the eve of the Cultural Revolution, it had risen to an estimated level of 10 million tons. By 1970 output had again doubled, and rapid increases were registered in subsequent years so that output in 1972 was close to, if not in excess of, 30 million tons.

Small exports to North Korea and North Vietnam probably took place after 1965. However, it was only in 1972 that Japanese businessmen began to discuss in earnest the possibility of oil import from China. The attractiveness of Chinese oil increased in Japanese eyes with the threatened shortage and the rising world price of oil, Japan's need for additional leverage in negotiations with the Soviet Union on the various Siberian projects, and the possibility of using oil as an area of cooperation with Peking *per se*. On the Chinese side, the rising oil price makes oil export a highly desirable addition in the Chinese balance of payments, especially in the light of the country's expanded import program since the end of the Cultural Revolution in 1969. Reorientation of Chinese foreign policy in order to develop a deterrent posture vis-à-vis the Soviet Union required the rapid expansion of certain categories of imports as a part of the effort, both short-run and long-run, to build up defense and to pursue selected lines of economic development. Peking may also believe that, directly or indirectly, oil can perhaps be used as leverage in negotiations with Japan, not only to promote the PRC's economic and defense interests in a direct sense, but also to damage the respective interests of the Soviet Union and of the Republic of China (ROC) on Taiwan. Given these circumstances, it was not surprising that oil became a center of negotiations between China and Japan at this time.

The immediate impetus to these negotiations was provided by the formation of the new Japanese government under Prime Minister Tanaka, who had come into office after a hard-fought campaign for party leadership in the LDP. One important factor that contributed to the defeat of Tanaka's rival, Fukuda, was undoubtedly the radical change of U.S. policy toward China in 1971 without prior consultation with Japan, and the subsequent defeat of the U.S.-initiated UN resolution, which Japan was persuaded to give its consent to cosponsor, that would have kept a seat for Taiwan in the General Assembly. These major setbacks to Japan's foreign policy took place while Fukuda was foreign minister under the premiership of Sato, Tanaka's immediate predecessor. Tanaka was, therefore, particularly anxious to promote "normalization" of relations with Peking, even at the expense of denouncing Japan's peace treaty with the Republic of China. He needed to present to the Japanese public and the world at large the image of a new Japan no longer tied to the apron strings of Washington. There was a strong incentive for many Japanese to show that Japan could even move ahead of Washington, contrary to if not in defiance of the latter's preferences. For this reason economic cooperation, in which increased export to the Chinese market and oil import from China would play an important role, would be complementary to overall policy. Since trade between the PRC and Japan had consistently been in the latter's favor, the addition of a new large item on China's export list might induce China to purchase more from Japan even if Peking were to insist on maintaining bilateral balance. From the point of view of Japanese refineries and oil importers, the shorter distance to low-sulphur Chinese oil from Ta-ch'ing as an alternative to Indonesian oil appeared especially attractive, although the high wax content of Ta-ch'ing oil is a problem.

Following a visit of leading Japanese businessmen to Peking in the summer of 1972, continuing discussions were held in both countries. A Japanese delegation that went to Peking in January 1973 was composed of members from Idemitsu Kosan, Daikyo Oil, Maruzen Oil, and Kyodo Oil, the syndicate behind International Oil which had been formed to negotiate and conduct crude oil import from China. Success of the mission was announced in April 1973 by Yosamatsu Matsubara, president of International Oil.[40] Under this agreement International Oil would import an annual volume of 1 million tons of Chinese crude at either FOB or CIF prices that would be determined at the time of shipment. This flexibility in price quotation was designed, apparently by agreement, to allow for the higher transportation costs by small Chinese tankers which the Chinese ports could handle and the need for Chinese oil to be competitive vis-à-vis the then prevailing world price—and doubtless also the price of Soviet oil.[41] This first major contract was negotiated at the time of parallel Japanese-Soviet negotiations on the Tyumen project. By negotiating with the Soviet Union and China simultaneously, Japan was able to score an initial success in obtaining oil from both communist suppliers, albeit only in trivial amounts. An initial shipment of 13,000 kiloliters of Ta-ch'ing oil arrived at the Idemitsu Hyogo refinery on May 21, 1973, a few weeks after the Soviet shipment, which was also consigned to Idemitsu.[42]

Since Ta-ch'ing oil had to be shipped either to Ch'ing-huang-tao or Dairen (Ta-lien) for loading on tankers, there was considerable international interest in Chinese pipeline construction. An oil mission from Canada was in Peking in April 1973 to discuss possible pipeline sales and, not surprisingly, the Japanese were interested in participating.[43] A technical mission from Japan Steel Tube reportedly was invited by the Chinese to consult on the type of oil pipes suitable for moving low-sulphur, high-viscosity oil, such as is produced at Ta-ch'ing. Among the topics of discussion were techniques for heating stations and pipelines for offshore oil. In September of the same year Ichiro Kimura,[44] vice president of International Oil and board chairman of the Kansai head office of the Japan International Trade Association, reported that a pipeline from Ta-ch'ing to Chin-chou—about 600 kilometers in length—had already been built, and that its extension to the shipping ports would be completed by the end of

1974.[45] At this time International Oil announced that it intended to negotiate for an increase in crude oil import from China to 5 million tons in 1974 and that the amount would in any event be at least 1 million tons. The larger import was expected to be made easier through port expansion to accommodate slightly larger tankers.[46]

The period between the summer of 1973 and early 1974 was marked by uncertainty about Soviet intentions regarding Tyumen and a more realistic assessment in Japan of the prospects of Soviet supply as a result of the protracted negotiations between the two countries. While Japan's import of Chinese oil in 1973 was below the total amount of Soviet oil imported in that year, the Japanese spared no effort in expanding supply from China and in broadcasting its intentions. In February 1974, Ichiro Kimura announced from Peking that in anticipation of the establishment of a second negotiating and importing channel for Chinese oil, International Oil would increase its import from China in 1974 to 1.5 million tons.[47]

A second channel for Chinese oil import came into being in March 1974 when the Japan-China Oil Import Council was established. Headed by Ryotaro Hasegawa, chairman of the board of Asian Oil, the council included in i's membership a number of oil companies and trading firms.[48] In August 1974 Hasegawa announced that the council members would import 350,000 tons of Chinese oil during the third quarter of 1974, and that total oil import from China in 1975 would be sharply higher: 3–4 million tons in the first half of the year and 10 million tons in the second half. Hasegawa was also most enthusiastic about the prospects of Chinese oil production. The estimates he gave were 50 million tons in 1973—which corresponds to most official and unofficial estimates, 70 million tons in 1974, and 100 million tons in 1975. The increased output estimated for 1975 was based on expectations of sharp increases at the Ta-kang and Sheng-li fields, although Ta-ch'ing would continue to be the main source of production and export. Hasegawa expressed his hope of importing Sheng-li oil beginning in 1975 and Ta-kang oil as of 1977. With technical aid from Japan, he expected Chinese production to be able to reach 400 million tons by 1980. Of this amount one-quarter might then be exported to Japan.[49]

We should bear in mind the prevailing circumstances when Hasegawa made these projections. Japan was still recovering from the shock of the Arab oil embargo and was faced with demanding Soviet proposals. Whatever may have been his realistic appraisal of China's propects, it was in the interest of his group and of Japan as a whole to create an impression of potentially huge supplies from willing Chinese sources.[50]

Following successful negotiations between a Chinese oil mission and the two Japanese import groups, it was announced in Tokyo in October

1974 that import of Chinese oil in 1974 would reach 4.5 million tons. Of this amount 3 million tons would be imported by International Oil as against the 1.5 million tons expected in February 1974, while another 1.5 million tons would be imported by Hasegawa's group. The 1.5 million tons of the Hasegawa group included 350,000 tons for the third quarter of 1974 and 850,000 tons for the fourth quarter. For 1975 the prospect of an import of 8 million tons from the PRC was established as the minimum. At the same time an import of 10 million tons in FY 1975 (April 1975 through March 1976) was said to be regarded by some Japanese business circles as a possibility. The same sources also spoke of a further rise of oil import from China to 20 million tons "before long."[51] Again one gathers the impression that while the likelihood of a larger future flow of Chinese oil supply was quite real, the prospect was more than a little exaggerated by interested parties.

Shortly following the report cited above, a substantially lower estimate of oil import from China in 1980 appeared and was attributed to Matsubara and Hasegawa. Sixty million tons were now said to be the level of possible import from China as against 100 million tons mentioned in speculations in August. Such an import volume was expected to be equivalent to 14 to 15 percent of Japan's anticipated requirement at that time.[52] The PRC was also said to be planning sea berths for tankers in order to facilitate export, as well as expansion of its production to 400 million tons—the same figure given in September by Hasegawa—by the end of the 1970s. The price of Chinese crude at this time was $12.80 a barrel, and was expected to drop; the competitive Indonesian price was $12.60.

A reduction in the demand for oil and the relatively successful Japanese effort, especially with the cooperation of Arab suppliers, to increase crude import resulted in a surfeit of oil in Japan toward the end of 1974. According to the *Nihon Keizai*, 1 million tons of Chinese crude that had already arrived in Japan at the beginning of December were not immediately unloaded.[53] Oil refineries and domestic distributors asked both International Oil and the Hasegawa group to cancel some of the imports they had recently contracted and not to carry over into 1975 what could not be imported in 1974. Declining demand, especially by power companies which burned Chinese crude, had brought about this request. Furthermore, the price of Chinese crude had dropped to $11.80 per barrel, although in the meantime the comparable Indonesian crude was quoted at $11.60. Japanese oil users obviously would prefer a cheaper oil. To be caught buying the higher-priced Chinese oil at the same time they were negotiating with Pertamina[54] would pose a special difficulty to the Japanese in their bargaining with the Indonesians. The Chinese, on the other hand, urged the Japanese buyers to take delivery of the oil already imported under contract. The Japanese side then decided to send a special delegation to Peking in order to explain away its unwillingness to receive the oil which it had sought so eagerly only a very short time ago.[55] Lack of storage facilities and declining domestic demand were offered as the compelling reasons.

While both factors were genuine, this reversal of position raised a serious question about Japanese policy. Is there a real advantage for Japan to try to secure supply through long-term contracts arranged by private businessmen? The latter are eager to minimize cost when supply appears plentiful. On the other hand, they put a premium on assurance of supply, and do not worry about the price when future supply threatens to be short. From the point of view of the Chinese supplier, such long-term contracts must be discounted if the foreign buyer may choose not to honor them in the future, and this would be disruptive of Chinese economic planning.

Several developments reported in early 1975 were a logical aftermath. First, the Japanese government through MITI now wished to substitute bilateral government agreement for importation by private firms in order to stabilize the flow of imported crude from China—and, by extension, the total flow of oil import.[56] Second, an agreement was reached with Peking through the effort of a special mission—headed by Yoshiro Inayama, board chairman of New Japan Steel, and including Matsubara, Kimura, and others—changing the means of payment for oil to a dollar basis from payment on the basis of the yuan price. As a result of an artificially low dollar-yuan rate set by Peking, payment in yuan raised the price of Chinese crude by as much as one dollar over the price of Minas oil from Indonesia.[57]

Finally, negotiators from the two countries succeeded in reaching agreement on the total volume of crude oil import from China under the 1974 contract. This amount was set at 4 million tons instead of 4.9 million tons,[58] which would have included 900,000 tons contracted earlier but now scratched. Further, the total amount contracted for 1975 was set at 8 million tons. This was the minimum amount mentioned during discussions with the Chinese oil mission in Tokyo.

Prospects of Chinese Oil

Analogous to the Soviet-Japanese discussion on the possible manufacture of LNG,[59] Japan proposed to construct for the PRC an LNG plant of 150,000 tons in annual capacity. Peking reportedly was interested in building larger plants but was unwilling to supply data on the size of natural gas deposits. The obstacles to this potential project resemble those that dogged bilateral negotiations between the Soviet Union and Japan. In addition to the insufficient data problem, in this case, Japan must also face competition from both France and West Germany.

Kumagai, the deputy director general of the Resources and Energy Agency, who visited Peking in early 1975, disclosed upon his return several interesting facts.[60] First, no medium- or long-term bilateral governmental agreements had been reached with Peking during his visit. Second, contrary to Japanese hopes, Peking was still unwilling to deviate from its long-standing policy of not considering joint ventures with Japan—or any other foreign country or foreign business interests—for the development of the country's oil resources. The PRC was, however, interested in importing foreign technology and equipment embodying new technology.

Kumagai also mentioned that the PRC's crude oil production in 1973 totaled a little over 50 million tons and that it was growing at approximately 20 percent a year. At this rate the 1974 output would be around 60 million tons and production would rise to around 170 million tons in 1980 instead of the 400-million-ton level optimistically mentioned in previous public announcements by the private Japanese importing firms.[61]

Among the latest Japanese speculations on the future prospects of (1) PRC oil production, (2) volume of export available, and (3) potential Chinese oil export earnings, two projections have been published by the Japan External Trade Organization (Jetro).[62] The estimates are based on specific assumptions about (1) production of energy from different sources in the 1974 base year, (2) growth rates in the individual energy supply sectors, (3) rates of growth of Chinese domestic consumption in individual consuming sectors, (4) stability of the numerical values of the energy-consumption coefficients, (5) PRC policy toward the export of oil as against that of other energy products, and (6) crude oil price.[63]

According to the Jetro estimates the first projection would result in Chinese crude oil export increasing from 8 million tons a year in 1975 to 12 million tons in 1979. Alternatively, assuming lower energy-consumption coefficients, the estimated crude oil export in Projection I would increase from about 11 million tons in 1975 to 26 million tons in 1979. In Projection II, the corresponding estimate of oil export would rise from 10 million tons in 1975 to 28 million tons in 1979 or, at lower energy-consumption coefficients, from 13 million in 1975 to 43 million tons in 1979. Thus the range of estimates for the two projections for 1975 are 8 to 13 million tons and for 1979, 12 to 43 million tons. In both projections a quantum jump would occur in Chinese oil export in 1980 and thereafter because of the assumed rates of production increase and the much lower assumed rates of increase in domestic demand.

Several factors should be borne in mind in evaluating these projections.

TABLE 11

JETRO PROJECTIONS OF PRC ENERGY DEMAND AND SUPPLY

(in 10^{12} BTU)

			SUPPLY			
Year	Coal	Oil		Natural Gas	Hydro-electric Power	Total Supply
Projection I						
1974	8,498	2,730	63.6[a]	857	169	12,560
1975	8,923	3,276	76.3[a]	1,028	193	13,720
1979	10,845	6,793	158.3[a]	2,132	325	20,290
1980	11,388	8,152	189.9[a]	2,559	391	22,570
1985	14,534	20,284	472.6[a]	6,368	714	42,050
Projection II						
1974	8,498	2,730	63.6[a]	857	169	12,560
1975	8,923	3,358	78.2[a]	1,028	193	13,812
1979	10,845	7,498	174.7[a]	2,132	325	20,995
1980	11,388	8,998	209.6[a]	2,559	371	23,410
1985	14,534	18,899	440.3[a]	6,368	714	40,666

			DEMAND			
Year	Demand for Energy	Supply of Energy	Surplus Energy	In Oil Equivalent (million metric tons)	Assumed Constant Crude Oil Price ($/BBL)	Exportable Volume (millions of $)
Projection I						
1974	12,293	12,560	267	6.2	13.5	616
1975	13,362	13,720	358	8.3	12.0	734
			(469)	(10.9)		(962)
1979	19,777	20,290	513	11.9	11.0	968
			(1,138)	(26.5)		(2,147)
1980	21,277	22,570	1,293	30.1	11.0	2,440
			(2,050)	(47.8)		(3,869)
1985	32,381	42,050	9,669	225.3	11.0	18,248
Projection II						
1974	12,293	12,560	267	6.2	13.5	616
1975	13,362	13,802	440	10.3	12.0	902
			(551)	(12.8)		(1,130)
1979	19,777	20,995	1,218	28.4	11.0	2,299
			(1,843)	(42.9)		(3,478)
1980	21,277	23,416	2,139	49.8	11.0	4,037
			(2,895)	(67.1)		(5,463)
1985	32,381	40,665	8,284	193.0	11.0	15,635

SOURCE: Masahiko Ebashi, *Jetro China Newsletter* (April 1975), pp. 22–26. The estimates are said to be based on "China's Energy Economy and Foreign Trade" by Masanobu Otsuka, in *Chugoko Keizai Kenkyu Geppo.* The figures in parentheses under "Demand" are based on the assumption that energy use for general consumption would remain at the 1974 level.

[a]Million metric tons.

First, the estimates are based on the assumed changes in production and consumption. Apart from uncertainty about the base-year figures, by their very nature the assumptions are likely to hold better in the short run than in the long run. Second, past Japanese estimates, especially for the long term, have tended to be on the more optimistic side. In estimates of 1973–74 the short-term projections have generally represented downward adjustments of earlier longer-term estimates. In short, the short-term estimates have been far more realistic than the long-term ones. The latter have been affected, perhaps inadvertently, by a desire to see higher figures of Chinese oil production and exportable surplus. Emphasis on a large potential has the benefit of whetting the Chinese appetite for development, especially if some Chinese are so predisposed themselves, and therefore of making the PRC more eager to negotiate for Japanese equipment and technology. This benefit would obtain regardless of the degree of confidence Japanese oil men and others actually have in their own estimates. For our purpose we need only to bear in mind two points: (1) It is advisable not to go beyond four or five years in using these projections. (2) Within this time period, even the highest estimate for 1979 would be only 43 million tons, or roughly in the order of 10 percent of Japan's projected oil import in that year.[64]

For any real large increase in production beyond the present level, the PRC will have to develop its offshore oil. Otherwise a period of long-term growth rate at 20 percent may not be sustainable. One Canadian author has suggested that the realistic long-term growth rate may be closer to 11–13 percent a year.[65] If so, the lower estimate of Jetro's Projection I (8 to 12 million tons a year of oil export) may prove to be closer to the mark.

Some Interim Conclusions

The preceding detailed account of Japanese negotiations with the Soviet Union and China during 1973–74 has revealed some interesting highlights which we can briefly summarize here. In doing so, we must bear in mind the strong underlying military and political considerations entertained by both sides of each of the two sets of simultaneous bilateral negotiations, as well as their interaction. Hence, any significant shift in the international environment or in the domestic politics of the three countries could bring about agreement where previously agreement was not possible, or disagreement where previously agreement was possible. One should not, therefore, be surprised if new departures and outcomes develop in the wake of the many policy adjustments that must take place following the collapse of the American effort in Vietnam in 1975. With this caveat in mind we can nonetheless advance a few general observations that appeared correct as of mid–1975.

First, Japan's intensive negotiations with the Soviet Union in 1973–74 resulted in no significant agreement that could assure Japan of a substantial amount of crude oil import from Soviet Siberia. With considerable uncertainty, the most that Japan could expect to receive from Soviet Far Eastern and Siberian sources would be a somewhat larger amount of Sakhalin oil at an annual rate of 5 million tons. The contribution of this source to Japan's total crude import in 1985 would be in the neighborhood of 1 percent. Because of the long lead times required in negotiation, exploration, and construction, one can safely rule out for the near to medium term—say, up to five years—any large expansion of Siberian crude oil import by Japan as an outcome of joint projects between the two countries. Such a development involving Siberian oil would have to result from Soviet developmental effort alone—together, of course, with appropriate changes in Soviet domestic production and consumption. A more probable increase in the supply of Soviet oil to Japan, if it occurs at all, would have to come from production in Western Soviet Union or through swapping via intermediaries. From the point of view of Japan, therefore, even under such conditions it would not be possible to increase the degree of Japan's diversification and direct control over oil supply by taking advantage of Soviet sources.

In trying to secure oil from the PRC, Japan was more successful during 1973–74. However, at the rate of 8 million tons a year, import from China would not account for more than some 2–3 percent of Japan's total crude import in 1975. In the Chinese case no less than in other cases, the rate of expansion of production in relation to that of China's domestic demand will govern the exportable surplus available if oil export is treated as the "residual." The determinants on the export supply side are in turn governed by Peking's attitude toward admitting foreign participation in developing oil resources, especially offshore, and the PRC's general economic policy on resource allocation. In regard to the last point, some of the unresolved questions consist of the evaluation of alternatives, such as (1) a possible reduction of energy export in order to manufacture more fertilizers which could increase domestic grain production and reduce net grain and fertilizer import; (2) allocation of more energy products to the production of synthetic fibers that would reduce domestic consumption of cotton textiles, thereby increasing textile export and/or making land available for more grain production and even export. Alternatively, if oil export is needed for balance of payments reasons, Peking may have to adjust either oil consumption or production, or both, in order to meet export goals. Deterioration in the PRC's terms of trade between imports and exports other than oil may even make such a development plausible. However, in such a case, Peking's future economic growth could be adversely affected. Since resolution of all these questions must be predicated upon certain political and ideological decisions, a sharp increase in China's oil export and production above the level mentioned by Kumagai in the near or medium term would again be unlikely. However, allowing for production increases at the existing major producing centers, beyond their current levels, especially Sheng-li and Ta-kang in addition to Ta-ch'ing, it may be possible to increase Japan's oil import from China substantially. Nevertheless one can safely take an arbitrarily high enough figure, such as 10 percent of Japan's total oil import in 1985, as an upper limit. Actual performance is likely to fall short of the limit by a wide margin.

If the effort to improve Japan's energy position by dealing with the Soviet Union and the PRC has thus far yielded relatively meager results, the underlying reasons deserve examination. One obvious reason was the lack of existing developed sources of supply with vast additional capacity that could be tapped immediately in both the Soviet Union and the PRC. However, in the Soviet case, Moscow's tactics in its negotiations with Japan were at least partly responsible for having so little to show in the end.

There is a fundamental difference between the Soviet Union and Japan in the considerations each side must take into account. For Japan the potential gain lies in a larger supply of oil from a nearby source, although even if this supply should become available, it could still be turned off unilaterally by the Soviet Union. Hence, the potential supply carries with it an irreducible amount of risk.

A major part of the cost to Japan in entering into joint ventures with the Soviet Union consists of the large initial investments required. It must be borne by private investors in the final analysis, however much governmental and even U.S. participation might have reduced both the risk and the cost. Another cost is the political and strategic net advantage that would be gained by the Soviet Union through (1) Siberian development and (2) increase in oil supply to the Soviet Far East. This cost must be evaluated within the framework of the trilateral balance involving the Soviet Union, the PRC, and Japan, not to mention the effect of any shift of the balance on the future role of the United States in this region. For Japan, the potential economic advantage—discounted for its inherent risk— must be evaluated against the two obvious costs. Costs here are both economic and noneconomic, and they must be separately assessed by private businesses and government agencies respectively. In the case of disagreement during the assessment process, caution probably takes over. Distrust of the Russians by many Japanese contributes to this balancing process.

For the Soviet Union, the cost of Japanese participation in Siberian development consists primarily of the oil exports that must subsequently be sold to Japan although there will be another quid pro quo at the time of sale. The noneconomic cost or disadvantage of having to disclose certain information to the Japanese—and, presumably, indirectly to the Americans —can also be worrisome. However, one would assume that, given normal internal security arrangements within the Soviet Union, this disadvantage can be contained. On the other hand, the advantages, both economic and noneconomic, are enormous. The economic advantage obviously lies in the possible acceleration of Siberian development through importation of Japanese resources and technology. During the period when Japanese credit has not yet been fully repaid, economic self-interest would enable the Soviet Union to hold Japan in virtual hostage, inasmuch as creditors are always at a disadvantage when they have no way to foreclose on a mortgage. The political and military advantage, even if only a pipeline were built to convey Tyumen oil to the Siberian coast, would be very large. In the circumstances, one wonders why Soviet negotiators would try to press their advantage to such a degree as to undermine Japan's obvious willingness to make accommodations. One gains the impression that in pressing the Japanese for additional concessions on the terms of the loan and the magnitude of the credit, or in introducing the Baikal-Amur railway proposal, the Russians overplayed their hand unnecessarily. Is this a miscalculation or is it poor coordination in Soviet planning between economic planners and the defense establishment? Was the prospect of a large exportable surplus of Soviet oil much dimmer than expected? If so, could not a promise have been made in any case?

Insofar as the PRC is concerned, the Chinese are of course in need of larger export income. During 1974, their greatly increased oil-export income from Japan was largely offset by sharply higher prices of industrial imports from Japan and reduction of Japanese imports of other Chinese goods.[66] The Chinese are also anxious to help strengthen Japan's bargaining position vis-à-vis the Soviet Union. They would like to divert Japan's potential oil purchase from the Soviet Union entirely to themselves. Hence, they too have an incentive to exaggerate their oil claims. In the final analysis, however, Peking's freedom of action is limited by the level of its physical output of oil. The prospects offered by offshore development remain uncertain as long as the country's internal political struggle and ideological debate preclude the admission of foreign technology and personnel on a large scale. Since these problems cannot be quickly resolved, the most that Japan can expect to accomplish in diversification by importing more crude oil from the two major Asian communist countries will remain limited in the second half of the 1970s. If less is imported from the Middle East, other non-Communist sources will have to bridge the gap.

4

EFFORT TO DEVELOP "AUTONOMOUS" SUPPLY

Direct Import of Oil by Japanese Trading Firms and Bilateral Governmental Arrangements

DD (direct deal) oil. In order to increase its sense of security and independence in the supply of imported oil, Japan has tried to increase the share of oil *under direct Japanese control*. This goal is approached in several ways. First, an effort has been made to import more oil through Japanese trading channels. Such oil is known as "direct deal" or "DD" oil. Oil obtained in this manner bypasses the major international oil companies, thus reducing the risk of diversion of supply away from Japan when there is a general oil shortage that also affects the home countries of the international companies. Some Japanese are clearly under the impression that such a diversion was actually perpetrated at their expense during the winter of 1973–74.

The disadvantages of DD oil are twofold. First, as long as the volume of direct Japanese purchases is small, it may be higher priced than oil from the international companies, partly because of the small bidders' natural disadvantage. Second, Japanese trading houses are notoriously competitive. When faced with severe shortages on international markets, they may well bid against one another, thus raising the price even further. During the first quarter of 1974, when the share of oil imported by international companies declined and DD oil—together with oil from Japanese-owned exploration companies—increased, the price of crude oil imported by Japanese trading and exploration firms was approximately 22 percent higher than that of the international "majors." This is equivalent to 4,000 yen per kiloliter or $2.12 per barrel.[1] Japanese refineries dependent upon this source of supply—since the international companies were assumed to supply their own downstream affiliates first, and probably did so—were therefore at a distinct cost disadvantage. A decrease in the market share of sales of refined products by Japanese firms was attributed to this factor by the national companies. Japanese industrial consumers have also pointed out that trading companies for which oil importation is only one of its many activities may be willing to pay a higher price for foreign crude oil if they can obtain a larger profit in their other commodity purchases or sales as a result of their oil purchase. A potential conflict of interest between the oil importer and the oil consumer therefore exists in the case of DD oil.

During FY 1968 the major international oil companies were responsible for about 72.2 percent of Japan's total crude oil import. This ratio stood at 74 percent in FY 1972 and 68.1 percent in FY 1973. The lower 1973 figure was doubtless an effect of events in the crisis months. U.S. firms led the list of the major oil companies, followed by the British and the French. The pattern of market shares can be seen in Table 12.

As of 1974 one Japanese trading company was involved in dealings with Abu Dhabi; six companies belonging to three financial groups were involved with Iraq; fourteen companies in eight financial groups were involved with Saudi Arabia. Altogether DD oil accounted for 12 percent of Japan's total crude oil import in March 1974 (about the same as the entire first quarter) against only 7 percent during the first half of 1973, before the oil crisis. The comparable figures for the major international oil companies were 65 percent in March 1974 against 70 percent in the first half of 1973.[2] These figures, which are for very short periods, are not

TABLE 12

CRUDE OIL IMPORTED INTO JAPAN BY MAJOR INTERNATIONAL
OIL COMPANIES

(percentage)

	FY 1972	FY 1973
U.S. Based Firms:		
Caltex	15.7	15.6
Exxon	11.9	11.3
Mobil	9.2	7.9
Getty	3.9	2.5
Unoco	3.4	2.8
Gulf	8.1	7.9
Other	3.0	2.8
Subtotal	55.2	50.7
British Based Firms:		
SIPC	12.8	10.9
BP	4.2	4.5
Subtotal	17.0	15.4
French — CFP	1.8	2.0
Total	74.0	68.1

SOURCE: *Data on Oil Development*, 1974, p. 5.

necessarily indicative of a definite trend, however, in spite of the special effort to expand DD supply as apprehension about a worldwide oil shortage spread.

GG (government-to-government) oil. The principal bilateral oil-import arrangement Japan had in 1974 was with Iran. As mentioned earlier, attempts to make similar arrangements with the PRC have thus far failed. The arrangement with Iran provides for Japanese import of 150 million tons of Iranian crude over a ten-year period. The annual amount is therefore small. The quid pro quo in this case is a billion-dollar credit to Iran for a petrochemical complex, a refinery, and a cement plant, plus technical aid. There is a similar arrangement with Iraq. Another potential candidate appears to be Libya. More GG possibilities may develop as more oil-exporting countries nationalize their oil and have larger quantities for direct sales by the government. A shift from DD to GG could also take place—as may happen in the case of Saudi Arabia—if the producing country prefers to deal with a larger government buyer than with many private trading companies.

Overseas Oil Exploration

The recent record (1967–74). If DD oil suffers from a cost disadavantage while GG arrangements are still limited, the third alternative to develop a secure supply under direct Japanese control is to engage in oil exploration and operation overseas through Japanese firms. This policy of increased autonomous supply through exploration and development began with the establishment of the Japanese-owned Arabian Oil Company in February 1958 for oil exploration and production in the Neutral Zone between Saudi Arabia and Kuwait. The first 1.5 million kiloliters of crude oil under this category of autonomous supply were imported into Japan in 1961. This was equivalent to 3.8 percent of Japan's total oil import during that year. By 1973 import of oil supplied by Japanese-owned overseas oil companies had risen to 24.6 million kiloliters. However, since the total volume of imported oil had risen even faster, the share of autonomous supply in 1973 was no more than 8.5 percent of the country's crude oil import.[3]

Compared to the volume of DD oil which accounted for 7 percent of total crude oil import in the first half of 1973,[4] oil from autonomous supply was responsible for about 8 percent during the same period. In March 1974 their respective shares were each 12 percent.

The scale of effort in overseas exploration for oil may be measured in

several ways. First, the number of Japanese firms engaged in oil exploration and/or production in non-Communist countries increased from a
single firm in 1968 to 49 in 1974. The 1974 figure includes several companies which had become dormant shortly after their opening, but
excludes the Sakhalin Oil Company which is interested in Soviet Sakhalin.
During the first eight years beginning in 1958 (1958–65), only four firms
were established (see Table 13). The rate of increase accelerated considerably thereafter. Three new firms then came on the scene in 1966, five in
1969, and from six to eight a year between 1970 and 1974 inclusive. The
formation of the Japan Petroleum Development Corporation (JPDC)
under MITI in 1967, to help finance overseas ventures by syndicates of
private interests through participation in risk capital and loan guarantees,
was a decisive step in promoting oil exploration abroad.[5] The year 1967
may therefore be regarded as a benchmark in the historical development
of Japan's oil policy. The appearance during 1969–74 of a number of companies formed by such major financial groups as Mitsubishi, Mitsui, and
Sumitomo, specializing in financing oil ventures and acting both as holding
companies and as direct operators of their own projects, further contributed to a speedup in expansion. By 1974 there were twenty-one firms
in Southeast Asia, nine in the Middle East, six in Africa, eight in South
America, four in North America, and three in Australia, including those
that had become dormant. Since some of these companies have ventures
in several countries while others have multiple ventures in a single country, the scale of the aggregate effort would be larger in terms of numbers
of projects. Table 14 provides an account of project numbers based on
information from JPDC.

Second, corresponding to the increase in the number of individual firms,
the level of expenditure on exploration and development rose sharply.
During 1958 and 1959 annual expenditures on overseas oil exploration by
Japan were 3.9 billion and 3.1 billion yen respectively. There was no
expenditure on development during these two initial years. Between 1965
and 1968, when the number of Japanese firms in overseas oil exploration
rose from four to ten, there was a simultaneous sharp increase in expenditures on exploration and development (see Table 15).

The above data on expenditures include some directly incurred by
JPDC. As of March 1974, the total request in the special budget for coal
and oil development for FY 1975, for example, included 38.2 billion yen
for oil development—including expenditures of JPDC, surveys for oil and
natural gas, data collection on overseas developments, and surveys or
studies on the environmental impact of explorations on the continental
shelf, on oil storage subsidies, and on oil movement. This amount exceeded
that of the preceding year by more than 10 percent.[6] Since some of the

TABLE 13

Growth in the Number of Japanese Oil Exploration and Development Companies by Region 1958–74

Year	Middle East	Southeast Asia	Australia (including New Guinea)	Africa	North America	South America	Sakhalin	Total excluding Sakhalin	Total excluding Sakhalin (net of duplication)
1958	1 (#1)	—	—	—	—	—	—	1	1
1959	—	—	—	—	—	—	—	—	—
1960	—	1 (#2)	—	—	—	—	—	1	1
1961	—	—	—	—	—	—	—	—	—
1962	—	—	—	—	—	—	—	—	—
1963	—	1 (#3)	—	—	—	—	—	1	1
1964	—	1 (#4)	—	—	—	—	—	1	1
1965	—	—	—	—	—	—	—	—	—
1966	—	1 (#5)	—	—	2 (#6, 7)	—	—	3	3
1967	—	1 (#8)	—	—	—	—	—	1	1
1968	2 (#9, 10)	—	—	—	—	—	—	2	2
1969	1 (#11)	3 (#12, 13, 14)	—	—	1 (#15)	—	—	5	5
1970	1 (#16)	1 (#19)	—	2 (#17, 18)	1 (#20)	1[a] (#21)	—	5, 1[a]	5, 1[a]
1971	2 (#22, 23)	4 (#25, 26, 27, 28)	1 (#29)	1 (#24)	—	—	—	8	8
1972	—	2 (#30, 31)	1[a] (#35)	1[b] (#32a), 1[a] (#33)	—	1[b] (#32), 1 (#34)	—	3, 2[a], 1[b], 1[b,c]	3, 2[a], 1[b]
1973	2 (#36, 37)	2 (#39, 40)	—	1[b] (#38)	—	3 (#41, 42, 43), 1[b] (#38a)	—	7, 2[b,c]	7, 1[b]
1974	—	4 (#44, 45, 47, 50)	1[c] (#46)	1[c] (#48)	—	1[c] (#48)	1[c] (#49)	4, 2[c]	4, 2[c]

Subtotal	9	21	1 1a 1c	3 1a 2b,c	4	4	4 1a 1b 1c 1b,c	8	1c
Total	9	21	3	6	4		51	1	49

Year Established	#	List of Companies
1958	1	Arabian Oil Co., Ltd. (ME)
1960	2	North Sumatra Oil Development Corporation (SEA)
1963	3	Kyokuto Petroleum Industries, Ltd. (SEA)
1964	4	Sabah Teiseki Oil Co., Ltd. (SEA)
1966	5	Japan Petroleum Exploration Co., Ltd. (SEA)
	6	Japan Canada Ltd. (NA)
	7	Alaskan Petroleum Development Co., Ltd. (NA)
1967	8	Kyushu Oil Development Co., Ltd. (SEA)
1968	9	Abu Dhabi Oil Co., Ltd. (ME)
	10	Middle East Oil Co., Ltd. (ME)
1969	11	Qatar Oil Co., Ltd. (ME)
	12	Mitsui Oil Exploration Co., Ltd. (SEA)
	13	Sabah Marine Areas, Ltd. (SEA)
	14	Sabah Oil Development Co., Ltd. (SEA)
	15	Maruzen of Alaska, Inc. (NA)
1970	16	United Petroleum Development Co., Ltd. (ME)
	17	Egyptian Petroleum Development Co., Ltd. (A)
	18	Zaire Petroleum Ltd. (A)
	19	Japan Low Sulfur Oil Co., Ltd. (SEA)
	20	Overseas Petroleum Corporation (NA)
1972	21	a Colombia Oil Co., Ltd. (SA)
	22	Iranian Petroleum Corporation (ME)
	23	Overseas (Middle East) Petroleum Corporation (ME)
	24	Nigeria Oil Co., Ltd. (A)
	25	Kaiyo Oil Co., Ltd. (SEA)
	26	Southeast Asia Petroleum Exploration Co., Ltd. (SEA)

Year Established	#	List of Companies
	27	Tonen National Resources Development Co., Ltd. (SEA)
	28	General Petroleum Development Co., Ltd. (SEA)
	29	Oceania Petroleum Development Co., Ltd. (Australia)
	30	Idemitsu Exploration Co. (Thailand) Ltd. (SEA)
	31	C Itoh Energy Development Co., Ltd. (SEA)
	32	b Mitsubishi Petroleum (SA)
	32a	b Mitsubishi Petroleum (A)
	33	Madagascar Oil Development Co., Ltd. (A)
	34	Andes Petroleum Co., Ltd. (SA)
	35	a JILD Co., Ltd. (Australia)
1973	36	Japan Oil Development Co., Ltd. (ME)
	37	Japan Iraq Petroleum Co., Ltd. (ME)
	38	b World Energy Development Co., Ltd. (A)
	38a	b,c World Energy Development Co., Ltd. (SA)
	39	Tokyo Oil Development Co., Ltd. (SEA)
	40	Sumatra Petroleum Co., Ltd. (SEA)
	41	Sumisho Peru Sekiyu (SA)
	42	Sumitomo Petroleum Development Co., Ltd. (SA)
	43	Fuyo Petroleum Development Corporation (SA)
1974	44	Central Energy Development Corporation, Ltd. (SEA)
	45	Arabian Oil Development Co., Ltd. (SEA)
	46	c PNG Petroleum Co., Ltd. (Australia)
	47	Bengal Oil Development Co., Ltd. (SEA)
	48	c Japan Peru Petroleum Co., Ltd. (SA)
	49	c Sakhalin Oil Development Co. (S)
	50	Teijin Malaysia Exploration Co., Ltd. (SEA)

Key:

ME = Middle East   A = Africa   SA = South America

SEA = Southeast Asia   NA = North America   S = Sakhalin

a Closed.   b Same company with project in different regions.   c Without investment information.

SOURCE: JPDC, *Outline of Japan Petroleum Development Corporation* (January 1975), and various publications of Sekiyu Kogyo Renmei (PPF) in 1974, plus interview data. The duplicate companies are included in regional totals, but not in the grand total.

TABLE 14

CUMULATIVE NUMBER OF PROJECTS OF JAPANESE OIL
COMPANIES BY REGION

Region	Financed by JPDC			Not Financed by JPDC		
	Ongoing	Discontinued	Total	Ongoing	Discontinued	Total
Southeast Asia	12	1	13	8	1	9
Middle East	5	3	8	1	—	1
Africa	3	2	5	3	—	3
Oceania	1	4	5	2	—	2
North America	1	2	3	2	—	2
South America	4	2	6	4	1	5
Total	26	14	40	20	2	22

SOURCE: *Outline of Japan Petroleum Development Corp.* (January 1975), p. 12.

TABLE 15

JAPANESE EXPENDITURES ON OVERSEAS EXPLORATION
AND DEVELOPMENT

(in billion current yen)

Fiscal Year	Exploration	Development	Total
1958	3.9	—	3.9
1959	3.1	—	3.1
1960	6.8	7.5	14.3
1961	—	30.7	30.7
1962	0.3	11.1	11.4
1963	2.1	6.0	8.2
1964	0.7	4.7	5.4
1965	2.1	4.4	6.4
1966	3.5	7.9	11.4
1967	5.0	6.8	11.8
1968	12.7	8.6	21.3
1969	13.5	13.2	26.7
1970	24.9	10.7	35.6
1971	26.7	16.7	43.4
1972	60.7	52.9	113.6
1973	82.6	48.6	131.2

SOURCE: Original data from MITI. Taken from PPF, *Data on Oil Development* (1974), p. 24.
Note: Totals may not equal the sum of components due to rounding.

TABLE 16

Value and Percentage Distribution of New Japanese Overseas Oil Investment by Region and by Year

(in million current yen of capitalization of firms)

Year	Middle East Amount	%	Africa Amount	%	Southeast Asia Amount	%	North America Amount	%	South America Amount	%	Australia Amount	%	Total Amount	%
1958	25,000	100	–		–		–		–		–		25,000	100
1959	–		–		–		–		–		–		–	
1960	–		–		2,000	100	–		–		–		2,000	100
1961	–		–		–		–		–		–		–	
1962	–		–		–		–		–		–		–	
1963	–		–		7,000	100	–		–		–		7,000	100
1964	–		–		3,271	100	–		–		–		3,271	100
1965	–		–		–		–		–		–		–	
1966	–		–		19,640	79	5,376	21	–		–		25,016	100
1967	–		–		5,000	100	–		–		–		5,000	100
1968	17,040	100	–		–		–		–		–		17,040	100
1969	10,400	44	6,080	25	13,282	56	90	neg.	–		–		23,772	100
1970	4,617	19	11,008	38	3,000	12	9,000	37	1,686	7	–		24,383	100
1971	12,890	45	–		3,308	12	–		–		1,388	5	28,594	100
1972	–		720	6	2,125	17	–		8,700	71	734	6	12,279	100
1973	37,700	68	4,000	7	4,440	8	–		9,120	17	–		55,260	100
1974	–		–		6,400	100	–		–		–		6,400	100
Total	107,647	46	21,808	9	69,466	30	14,466	6	19,506	8	2,122	1	235,015	100

SOURCE: Same as Table 13. Based on the capitalization figures of 45 firms only.

TABLE 17

Cumulative Value and Percentage Distribution of Japanese Overseas Oil Investment by Region and by Year

(in million current yen of capitalization of firms)

Year	Middle East Amount	%	Africa Amount	%	Southeast Asia Amount	%	North America Amount	%	South America Amount	%	Australia Amount	%	Total Amount	%
1958	25,000	100	—	—	—		—		—		—		25,000	100
1959	25,000	100	—	—	—		—		—		—		25,000	100
1960	25,000	93	—	—	2,000	7	—		—		—		27,000	100
1961	25,000	93	—	—	2,000	7	—		—		—		27,000	100
1962	25,000	93	—	—	2,000	7	—		—		—		27,000	100
1963	25,000	74	—	—	9,000	26	—		—		—		34,000	100
1964	25,000	67	—	—	12,271	33	—		—		—		37,271	100
1965	25,000	67	—	—	12,271	33	—		—		—		37,271	100
1966	25,000	40	—	—	31,911	51	5,376	9	—		—		62,287	100
1967	25,000	37	—	—	36,911	55	5,376	8	—		—		67,287	100
1968	42,040	50	—	—	36,911	44	5,376	6	—		—		84,327	100
1969	52,440	49	—	—	50,193	46	5,466	5	—		—		108,099	100
1970	57,057	43	6,080	5	53,193	40	14,466	11	1,686	1	—		132,482	100
1971	69,947	43	17,088	11	56,501	35	14,466	9	1,686	1	1,388	1	161,076	100
1972	69,947	40	17,808	10	58,626	34	14,466	9	10,386	6	2,122	1	173,355	100
1973	107,647	47	21,808	10	63,066	28	14,466	6	19,506	8	2,122	1	228,615	100
1974	107,647	46	21,808	9	69,466	30	14,466	6	19,506	8	2,122	1	235,015	100

SOURCE: Same as Table 13.

Japanese oil projects are joint ventures with non-Japanese partners, total expenditures on individual projects benefiting Japan are greater than those incurred by the Japanese alone.

For comparison, Japanese expenditures on exploration on the continental shelf of Japan in FY 1972 amounted to 7.9 billion yen. Similar expenditures for offshore Japan were only 220 million yen in FY 1973. These figures contrast with 60.7 billion yen for exploration overseas in FY 1972 and 82.6 billion yen in FY 1973.[7]

Comparison can also be made between Japanese expenditures on oil exploration and corresponding expenditures by other countries. For example, in 1971 West Germany and Italy had a total oil consumption equal to that of Japan while Italian expenditures alone on exploration were 1.8 times that of Japan.[8] The increase in Japanese expenditures is a more recent phenomenon.

Still another measure of the growth of Japan's overseas exploration effort is the level of investment as reflected in the total capitalization of the firms involved. Beginning with the 25 billion yen of the Arabian Oil Company, established in 1958, the aggregate capitalization of the Japanese firms—of which some financial data are available in the case of forty-five companies—totaled 37.3 billion yen in 1965, 67.3 billion yen in 1967 when the JPDC came into being, and 235 billion yen in 1974. The largest amount of capital investment in a single year apparently took place in 1973. However, allowing for price inflation, the total capitalization of the Arabian Oil Company still dwarfs subsequent annual investments (see Tables 16 and 17).

The JPDC generally finances about 50 percent of the total expenditures of the individual projects it supports. The remainder is supplied by private firms. The volume of investments and loans made by JPDC totalled 4.1 billion yen and 1 billion yen respectively during FY 1968, its first full year of operation. In FY 1973 the corresponding figures were 20.3 billion yen in investments and 7.7 billion yen in loans. The sharpest increase took place between FY 1971 and FY 1972, and again between FY 1973 and FY 1974 (see Table 18). These amounts are of course still quite small in dollars.

Finally, the expanding scale of the Japanese effort may be seen in the number of countries where Japanese oil firms are active. This number rose from a single one in 1958 to two in 1960, three in 1964, five in 1966, and six in 1968. Thereafter, following the formation of JPDC, the rate of increase rose by at least two countries a year. By 1974 twenty-two oil-producing countries had played host to Japanese oil projects.[9]

Effort and result. The preceding description of several indicators of

TABLE 18

INVESTMENTS AND LOANS MADE BY JPDC

(in billion current yen)

FY	Investments	Loans	Total	Guarantees
1967	0.8	—	0.8	—
1968	4.1	1.0	5.1	5.6
1969	5.1	1.6	6.7	1.8
1970	8.6	2.1	10.7	3.9
1971	13.0	0.5	13.5	10.5
1972	19.0	3.9	22.9	17.4
1973	20.3	7.7	28.0	13.6
1974 (est.)	32.9	44.1	77.0	51.7

SOURCE: *Outline of Japan Petroleum Development Corp.*, (Jan. 1975), pp. 3–4.

TABLE 19

MEASURES OF EFFORT OR INPUT IN JAPAN'S OVERSEAS
OIL EXPLORATION

Effort or Input	1958	1967	1974
Capitalization of Japanese overseas oil firms (billion yen)	25.0	67.3	235.0
Cumulative numbers of Japanese firms in overseas exploration	1	8	49
Number of JPDC-financed projects[a]	—	1	7
Total number of projects[a]	—	1	11
Number of wells drilled[b]	—	44	180 (1973)
Expenditures on exploration and development (billion yen)	3.9	11.8	131.2 (1973)
Number of countries where Japanese oil projects are located	1[c]	5	22

[a]JPDC, January 1975.
[b]Taken from *Data on Oil Development*, 1974, p. 24.
[c]The Neutral Zone is counted here as a single entity.

growth can now be summarized so that a comparison can be made between effort and result. In addition to (a) the number of firms engaged in overseas operation, (b) their capitalization, and (c) expenditures on exploration and development, we can add (d) the number of wells drilled, and (e) the number of countries where Japanese oil projects are located as measures of effort. The last indicator can also be used to measure diversification by source.

As for the result of this effort, an obvious indicator is the amount of oil derived from Japanese-owned overseas sources and imported into Japan. In million kiloliters, this import was 15.8 in 1967 and 24.6 in 1973, or under 10 percent of total crude import in the latter year.

Several observations can be made on the basis of these figures. First, the growth in expenditures, in number of firms, in investment in these firms, and in the expansion of coverage of countries by them took place primarily after 1967 when the JPDC was established. Second, most of the firms established after 1967—apart from the holding companies whose primary function was to finance projects by other corporate entities—are small organizations each engaged in a few exploration projects only. In contrast, only a few companies have thus far become sizable producers.[10] Japan's policy to increase national control over its oil supply, as implemented by JPDC financing and projects, has resulted in a proliferation of small exploratory efforts that so far have had relatively little to show. Not insignificant is the fact that of the forty projects financed by JPDC during the 1967–74 period, fourteen were later terminated. On the other hand, only two out of the twenty-two projects not financed by JPDC were discontinued. The much higher failure rate of JPDC projects would seem to suggest the very marginal nature of some of them. This raises questions about judgment and technical performance, not to mention the detrimental effect on both of the sense of urgency and pressure under which Japanese policymakers and those who implement policy have had to work.

Perhaps not enough time has as yet elapsed to allow for a large increase in production. Of the remaining twenty-six JPDC projects in 1974, (excluding those already terminated), sixteen were still at the exploration or pre-exploration stage, four were being developed, and three were in production. Another three were under study for termination or the possibility that they might be farmed-out. Of the twenty remaining projects in 1974 that were privately financed, seventeen were at the exploration or pre-exploration stage and three were in production. The fact that Japan is a late-comer in overseas oil exploration and production explains in part the small visible accomplishments in terms of Japanese import of oil from national enterprises. The inevitable lead time required for exploration and development and the fact that the obviously rich deposits had already been bespoken before Japan arrived on the scene are among the important reasons. The situation could, of course, change. However, if any change is to take place, it can occur only in the 1980s at the earliest. Economic recession in 1974–75 further reduced the ability and willingness of private business to finance heavy exploration and development expenditures so that JPDC may find it necessary to promote amalgamation or even to form a single enterprise to carry on the program.[11]

Vulnerability and Diversification Assessed

We can now look at the extent to which the vulnerability of Japan's oil supply at source has changed as a result of the promotion of oil import from Japanese-owned enterprises abroad, direct purchases by Japanese oil-importing firms, and bilateral governmental arrangements between Japan and noncommunist oil-producing countries. Vulnerability at any particular source (or sources) can be measured by its (or their) share of Japan's oil import. Where more than one national source is involved, the existence of some cohesive factor that might galvanize the individual countries into joint action must exist before Japan would experience any vulnerability or threat from the group. Since political factors that could cause a group of oil-exporting countries to coalesce may change over time, both the source of threat and Japan's vulnerability may also change unexpectedly. With this uncertainty in mind, we can nevertheless note certain very limited improvements in the relative position of Japan.

First, while the Arab countries accounted for 55.3 percent of Japan's total crude oil import in 1967, their share fell to 46.6 percent in 1973 (see Table 20). Second, while the share of the Middle East—the Arab countries plus Iran—was 91.2 percent in 1967, it dropped to 77.6 percent in 1973. Third, insofar as the communist oil-exporting countries are concerned, the share of the Soviet Union and the PRC in Japan's direct oil imports—excluding any swapping through third-party intermediaries—remained relatively unimportant during this period, varying from 1.4 percent in 1967 to 1.1 in 1973. As pointed out in an earlier section, there was a substantial increase in Japanese oil import from the PRC in 1974, but its proportion in Japan's total oil import remained insignificant.

If we examine the relative importance of single national suppliers in Japan's oil import, the largest supplier in 1967 was Iran, followed by Saudi Arabia and Kuwait. In 1973 Iran's share had fallen to 31 percent from 35.9 percent in 1967. The second largest single source in 1973 continued to be Saudi Arabia. However, the third place was now taken by Indonesia.

Conceivably, supply of oil can also be disrupted in the course of its shipment. Hence, a distinction of different suppliers can be usefully made on the basis of the world's shipping lanes. In this connection, during 1967, 91.4 percent of Japan's oil import was derived from sources west of the Malacca Straits and only 6.7 percent from south of Japan and east of Malacca. In 1973 these shares had become 80.3 percent from west of the Malacca Straits and 18.4 percent from south of Japan and east of Malacca. Among the Middle Eastern suppliers, the proportion of those

TABLE 20

ALTERNATIVE MEASURES OF VULNERABILITY

(percentage)

Measure of Vulnerability at Source	1967	1972	1973
Import from Arab countries	55.3	43.4	46.6
Import from the Middle East	91.2	80.7	77.6
Import from the USSR[a]	1.4	0.2	0.5
Import from the PRC	—	—	0.6
Total from the two communist countries	1.4	0.2	1.1
Import from west of the Malacca Straits	91.4	83.1	80.3
Import from west of the Hormuth Strait (Persian Gulf)[b]	89.1	80.6	77.3
Import from south of Japan and east of Malacca	6.7	16.4	18.4
Import from:			
1st largest supplier—Iran	35.9	37.3	31.0
2nd largest supplier—Saudi Arabia	18.2[c]	16.7[d]	19.9[e]
3rd largest supplier—(1967 Kuwait; 1972–73 Indonesia)	17.5[c]	16.3	13.6
Degree of diversification:			
Smallest number of countries supplying 20 percent	1 (Iran)	1 (Iran)	1 (Iran)
Smallest number of countries supplying 50 percent	2[f]	2[g]	2[h]

SOURCE: Compiled from information given in Table 6.

[a] Direct import only, excluding swap.

[b] Obtained by subtracting Iraq from the figure under "Middle East" in Table 6 because of the pipeline to the Mediterranean. Amount of oil from Saudi Arabia dependent upon transit through the Persian Gulf would be lower to the extent the Trans-Arabian Pipeline might be used.

[c] If the Neutral Zone (14.1%) is added to Saudi Arabia and Kuwait, the total would be 49.8%.

[d] If one-half of the Neutral Zone (4.2%) is added to Saudi Arabia, the total would be 20.9%.

[e] If one-half of the Neutral Zone (2.7% is added to Saudi Arabia, the total would be 22.6%.

[f] Add Saudi Arabia or Kuwait or Neutral Zone.

[g] Add Saudi Arabia or Indonesia or Kuwait.

[h] Add Saudi Arabia; alternatively, adding Indonesia would reach almost 45%.

using the Strait of Hormuth at the mouth of the Persian Gulf also declined from 89.1 percent in 1967 to 77.3 percent in 1973.

Unfortunately for Japan, the general improvements noted above are insufficient to enable Japanese planners to ignore the following facts: First, both the Arab countries and the Middle East continue to be responsible for too large a share of Japan's oil import. Second, several individual countries such as Iran, Saudi Arabia, and more recently Indonesia, constitute very large single sources of Japan's oil supply. Obviously, their goodwill toward Japan as well as their self-interest must be sufficiently assuring if Japanese vulnerability is to be minimized. Third, the Malacca Straits, together with several others, continue to be major choke points in the oil movement to Japan, not to mention the inescapable vulnerability of large tankers at sea throughout the entire sea lanes.

The degree of diversification achieved so far is still quite minimal, as can be seen from Table 20. Diversification can, of course, be measured in a number of ways. For instance, we can denote the degree of diversification by the smallest number of oil-exporting countries that in the aggregate account for x percent of Japan's crude oil import, and we can set the value of x at some arbitrary level so that denial of the supply would produce a substantial—perhaps even an intolerable—impact on the Japanese economy. Thus, in Table 20, if x is set at 20 percent, the smallest number of countries accounting for 20 percent of Japan's crude oil import in 1967 would be one. If x is set at 50 percent, the smallest number of countries in 1967 would be two. Table 20 shows that these same numbers would still obtain in 1973. While the diversification program of Japan has increased the number of sources of supply, unfortunately too many of the new suppliers are still too small—and a few of them far too large.

Thus, diversification must be pushed further; from the Japanese standpoint it would be preferable to have many small suppliers increase their production and become larger oil suppliers to Japan. Geographically they would best be located east of Malacca and south of Japan. If these criteria are applied, the direction of future Japanese effort in diversification points to such countries as Thailand, Malaysia, Singapore, Australia, and the Philippines, both on land and offshore, as well as offshore Korea and Taiwan. A second preferred area would be North and South America; then would follow Africa and the North Sea. Finally, the Soviet Union and the PRC, possibly even the Indochina area now under communist control, would constitute a preferred area for diversification from the Japanese point of view.

The Soviet Union and/or the PRC, by virtue of their still very low shares in Japan's total oil import, are in fact prime candidates in Japan's diversification program. As discussed before, the prospects at both sources are not particularly promising, however, and increasing their shares would in turn increase the two communist countries' political leverage over Japan, as distinct from any consideration of their relative importance to Japan purely as oil suppliers.

Problems of Internal Consistency between
Diversification and National Control

In view of Japan's recent experience in the oil embargo and the price hikes by OPEC, the valiant effort of the Japanese government and oil firms in overseas exploration and production has proved insufficient because of the still very large shares of crude oil import enjoyed by the Middle East and the Arab countries. At the same time, Japan's effort in expanding the supply of DD oil during 1973–74, while successful in reducing the share imported through the major international oil companies, actually increased reliance on these two areas. In the case of GG oil, its further expansion would have the effect of increasing the share of Iran and Iraq and, therefore, of the Middle East. Thus, in order to lessen Japan's dependence on the international majors, the promotion of DD and GG oil has actually worked against diversification by source and in favor of concentration. The impact is equally unfavorable from the locational standpoint.

All this can be explained away by pointing to the short-term, paramount need of assuring adequate supply during and immediately after the 1973–74 crisis. The point would take on greater significance if, in the long run, greater independence from the international majors would always imply greater dependence on Arab and Middle Eastern suppliers. One interesting question is: Which oil-exporting countries are likely to have large quantities of exportable oil available for direct sale outside the distributive channels of the international majors and/or for direct bilateral governmental barters with such oil-importing countries as Japan? *Prima facie*, the answer would point to those countries that have national oil monopolies or high participation rates in production-sharing with the international majors. Outside the communist countries and Indonesia, the Middle Eastern suppliers—including Iran, and such North African suppliers as Libya—would belong to this category. Thus there may be an inherent dilemma between the objective of geographical and source diversification, on the one hand, and diversification of venues of supply, on the other.

This dilemma can be resolved only if the total amount of oil import from the Middle Eastern and/or Arab countries can be reduced at a faster rate than the volume of DD or GG oil supplied by them is increased. If Japan's total oil import is at a fairly stable or rising level, such a change in both sources and venues can occur only if diversification can be accelerated in favor of the present smaller suppliers and away from the locations west of Malacca. Also, more DD and GG oil might be secured from suppliers east of the Straits.

There is a plausible rejoinder to the above argument. Some oil-exporting countries, on their part, may find it distinctly advantageous to diversify their markets and the external distributive channels they use. Therefore, it is in their interest to have Japan—which represents a large market—bid for their oil side by side with Western European and U.S. buyers. It is also in their interest to have Japanese trading companies bidding against the international majors. Of course, it is equally in their interest to have individual Japanese companies bidding against one another.

Looking back at the oil crisis with the benefit of hindsight, one can argue that in the long run the Arab countries would not really wish to curtail supplies to Japan significantly, and that they never intended to do so even during the crisis, except that a credible threat of reduction was necessary in order to accomplish two paramount OAPEC purposes: a sharp increase in crude prices and the establishment of a diplomatic framework to exert pressure on the United States. Japan's vulnerability with respect to oil supply made its selection as a prime target of the embargo especially appropriate. The alacrity and nervousness with which Japan responded to the threat more than amply fulfilled Arab expectations.

The Middle East as a whole exported 932.5 million tons of crude oil and 56.5 million tons of distillates during 1973. Nearly 22 percent of the aggregate of 989 million tons went to Japan.[12] Thus Japan is a valued customer and as such is needed by the oil-exporting countries. Japan is also a useful balancing factor in the oil-exporting countries' negotiations with other oil-importing countries and with the major international oil companies. This argument of Arab and Middle Eastern self-interest, however plausible and valid, does not, of course, remove the remote possibility that an actual cutoff of supply to Japan could in fact take place, at any rate in the short run. Nor would the mere threat of such a cutoff, which can be equally disruptive to the Japanese economy, be totally dismissed. Consequently, how seriously one should regard the dilemma noted earlier is a matter for the Japanese government and people to judge. Their decision would depend upon the degree of prudence they choose to exercise.

A related idea was advanced by a few Japanese oil men interviewed in early 1975. Because of their desire to keep the Japanese market, some Middle Eastern countries may be quite receptive to the formation of Japanese firms or the expansion of existing ones in their territories. In some countries, like Indonesia, where ownership of all the oil that may be found in the future is already vested in the state, no Japanese company can actually *own* the oil it produces, but there is always production sharing. One might add that for small political entities such as Abu Dhabi, their bargaining power and their present level of sophistication would favor any alert foreign firm that has gained entry.

It is not at all clear that nominal ownership is a prerequisite of control over the mode of disposal of the product, a control which is needed to guarantee security of supply. The existence at any time of a realistic possibility of nationalization is sufficient to blur the legal distinction between ownership and control. The gradual extension of the ratio of participation by the host country in the amount of oil produced, leading ultimately to 100 percent participation—which is equivalent to nationalization insofar as future production is concerned, has already gone a long way in Middle Eastern countries. The most important recent case is the expected final nationalization of Aramco by Saudi Arabia, even though Aramco's managerial role has actually expanded. Similar extensions of participation—up to 60 or 75 percent of production—have been and are being made by others, and further progression in this direction can be expected. Abu Dhabi, on which Japan has focused a major part of its interest, is a good example; so far it has followed the footsteps of Saudi Arabia.

Another potential internal inconsistency in Japan's program to develop overseas oil projects under Japanese control concerns the role of foreign-based major oil companies and independents. Many of the new Japanese oil firms and individual projects established in recent years have the participation of third parties other than the governments or private organizations of the host countries. Some joint ventures are the outcome of the introduction of a Japanese interest into an originally non-Japanese venture. Where the initiative has come from the Japanese side, there are several possible reasons. Japan still lags behind some Western countries— especially the United States—in oil technology, and there is, in particular, a shortage of young, adequately trained petroleum engineers. Some Japanese firms prefer to have third parties front for them in dealing with host governments, either because of their own unfamiliarity with local laws and conditions or because they believe that hostility toward Japan may still be felt by the local government or population. The last consideration applies especially in Southeast Asia. From the standpoint of both the Japanese and the third parties, financial participation by Japan—which has built up large foreign assets through accumulated export surpluses— can be quite welcome if they happen to have heavy demand for capital in other priority areas, such as investment in North Sea oil exploration.

Table 21 lists most of the major non-Japanese oil companies involved

TABLE 21

PARTICIPATION OF WESTERN OIL COMPANIES IN JAPAN'S
OVERSEAS OIL EXPLORATION

1974

	Participating non-Japanese Companies	Middle East	Southeast Asia	Africa	North America	South America	Australia	Total Number
USA	Sun Oil	1	1	1	–	1	1	5
	Mobil Oil & Gas	1	2	–	–	–	–	3
	Cities Service	1	–	–	1	1	–	3
	Continental	–	1	–	–	–	2	3
	Ocean Venture	–	–	–	–	1	2	3
	Union Oil	–	1	–	1	–	–	2
	Stanvac	–	2	–	–	–	–	2
	Ashland Oil	–	2	–	–	–	–	2
	Gulf Oil	–	–	1	1	–	–	2
	Getty Oil	–	–	–	1	1	–	2
	Signal Oil	–	–	–	1	1	–	2
	Occidental Petroleum	–	–	–	–	1	–	1
	IIAPCO	–	1	–	–	–	–	1
	Amoco	–	1	–	–	–	–	1
	Trend	–	1	–	–	–	–	1
	Skelly Oil	–	–	–	1	–	1	1
	Pan Ocean	–	–	–	–	1	–	1
	Trans World	–	–	–	–	1	–	1
	Superior	–	–	–	–	1	–	1
	Cayman	–	–	–	–	1	–	1
	Reserve Oil	–	–	–	–	1	–	1
	City Investing	–	–	–	–	1	–	1
UK	British Petroleum	1	1	–	–	–	1	3
France	CFP	1	1	–	–	1	–	3
	Aquitaine	–	1	–	–	–	–	1
	FRAP	1	–	1	–	–	–	2
Germany	Deminex	–	1	–	–	1	–	2
Italy	AGIP	–	–	1	–	1	–	2
Belgium	Cometra	–	–	1	–	–	–	1
Canada	Bour Valley	–	1	–	–	–	–	1
	Union Oil	–	–	–	1	–	–	1
	Total	6	17	5	7	15	6	56

	Middle East	Southeast Asia	Africa	North America	South America	Australia	Total Number	%
				Distribution of Companies				%
USA	3	12	2	6	12	5	40	71.4
UK	1	1	–	–	–	1	3	5.3
France	2	2	1	–	1	–	6	10.7
Belgium	–	–	1	–	–	–	1	1.8
Germany	–	1	–	–	1	–	2	3.6
Italy	–	–	1	–	1	–	2	3.6
Canada	–	1	–	1	–	–	2	3.6
Total	6	17	5	7	15	6	56	100.0

SOURCE: Same as Table 13.

in Japan's overseas oil firms and projects. The predominance of U.S.-based oil firms is unmistakable. Of course, cooperation with U.S. firms in the Japanese oil industry is a matter of long standing. A number of Japanese firms are undertaking or have undertaken offshore exploration around Japan in cooperation with foreign companies—e.g., Japan Petroleum with Caltex, Teikoku Oil with Gulf and Esso, Idemitsu with AMOCO, and Nishinihon with Shell. However, when such joint ventures are carried on in other countries where Japanese law cannot prevail, how much control will be vested in Japanese and how much in non-Japanese hands becomes a moot question. Is the development of autonomous supply under complete Japanese control a long-term goal while short-term cooperation with the major international oil companies and other foreign firms is an

TABLE 22

ESTIMATED PERCENTAGE DISTRIBUTION OF SOURCES OF FINANCING OF 45 JAPANESE OVERSEAS OIL FIRMS

JPDC	26.8
Oil production	18.8
Oil refining	9.0
Trading companies	12.6
Electric utilities	5.0
Metallurgy and metals	4.0
Shipbuilding and machinery	1.4
Banks	3.2
Chemicals	1.6
Insurance	2.2
Electric machinery	0.6
Gas companies	0.6
Manufacturing	0.4
Shipping	0.8
Textiles	2.0
Transportation	neg.
Brokers	0.8
Governments	2.2
Other	8.4
Total	100.0

SOURCE: Same as Table 13.

unavoidable tactic of national policy? Or have private Japanese oil companies reintroduced interdependence between Japan and the third parties,
not to mention the host countries themselves, in their drive toward economic independence under MITI's administrative guidance? If a real
internal inconsistency exists, has it arisen unknowingly?

The same questions can be raised from the point of view of non-
Japanese oil companies now cooperating with Japan in the latter's search
for oil. Are they looking for a degree of long-term cooperation and interdependence with Japanese interests? Or are they being used by Japanese
interests that have a present need for their technical, financial, managerial,
or—within a particular host country—special political competence? Or
might the answers to these questions be really quite immaterial to them?
Finally, would the governments of the countries in which these non-
Japanese firms are based regard the matter in the same way as the private
companies?

Still another area of potential inconsistency is the possible conflict of
interest between the oil-consuming firms and the trading companies, both
of which have been deeply involved in the overseas search for oil. As we
have already pointed out, the trading companies have a much broader
interest than the oil users. To them, trading in oil *per se* does not have to
be particularly profitable as long as it can lead to other profitable business.
This attitude not only could lead to their making higher bids in acquiring
DD oil, but it also could affect their counsel as part owners of ventures in
overseas oil exploration and production. Table 22 shows the extent of
direct participation by trading companies in supplying venture capital to
be about 13 percent in forty-five of the oil firms engaged in such activities.

In contrast to the preceding factors which point to potential inconsistencies and pitfalls, another consideration pointing in the opposite direction deserves at least brief mention. This concerns the involvement of
Japan's leading financial groups in the development abroad of oil resources
and the geographical distribution of their ownership interests.

As we examine the equity ownership of thirty-one firms involved in
overseas oil ventures where information is readily available, we can estimate for nine major financial groups their individual stockholdings.[13]
Since some of the stockholders are corporations that are in turn owned by
some of the same financial groups, indirect ownership once or even twice
removed can also be traced. In this manner we can estimate aggregate
ownership interests—and presumably influence—of nine major Japanese
financial groups in Japan's oil industry abroad. Since the available data
are incomplete and errors and omissions are unavoidable, the present
findings are necessarily very tentative. With these caveats in mind, the
following observations appear warranted:

First, if the world outside the communist countries and Western Europe is divided into five regions—the Middle East, Southeast Asia and Oceania, Africa, North America, and Latin America—the nine major financial groups can be ranked in terms of their ownership interest and geographical coverage roughly as shown in the table below.

Rank	Number of Regions
1. The Sanwa Group	4
2. The Mitsubishi Group	4
3. The Sumitomo Group	4
4. The Mitsui Group	3
The Fuyo Group	3
The DKB Group	2
5. The Toshiba-IHI Group	1
The Nissan Group	1
The Toyota Group	1

Second, the region with the largest number of groups represented in it is Southeast Asia (eight groups) followed by Latin America (five), Africa (four), the Middle East (three), and North America (one).

Bearing in mind again that our data are incomplete, these figures nevertheless suggest that there is a heavy concentration of interest in Southeast Asia and Latin America, but that the larger financial groups are themselves highly diversified geographically.

These points would seem to indicate a substantial convergence of interest between the large financial and industrial groupings of Japan and government policy as represented by MITI and JPDC under it. That is to say, the policy objectives of diversification and autonomous control of supply have the broad support of the financial groups even if there are certain serious internal inconsistencies and possible conflicts of interest. However, this does not mean that the degree of diversification and autonomous control of supply attainable in the long run will definitely be high.

Prospects for Japanese Investment in Foreign Oil

Given the small beginning Japan has made so far in its overseas oil investment, whether the objective of geographical diversification can be attained in the long run will depend upon the degree of Japan's success in finding oil especially in countries in Southeast Asia and Oceania, as well as in Latin America. The objective of increasing autonomous supply from these areas can be attained only if Japan can consolidate its access to these sources through ownership, management and operations contracts, or some other arrangement. In any event, Japan as a nation and Japanese businessmen who will become personally involved must enjoy the goodwill of the host countries in the prospective areas. Furthermore, assuming the presence of goodwill, these countries must not fall into such a position that they can be compelled by third nations to become outright hostile to Japan or to close their doors to Japanese business.

Mutual perceptions by Japanese investors and host countries. Consider first the issue of goodwill which must be entertained by both sides if it is to last. How do some of the oil-exporting countries regard Japan and Japanese businessmen? Conversely, what do Japanese oilmen, and Japanese businessmen in general, think of some of the host countries where they operate? During research trips to several Southeast Asian countries in 1973–75, a number of interviews conducted by this author with local officials and businessmen as well as with some third-country nationals have led to certain impressions that generally are confirmed by Japanese observers themselves.

In the report of a 1974 discussion at which various Japanese oilmen with considerable experience in overseas operations exchanged views, one executive from Abu Dhabi Oil observed that it was important to distinguish among (1) "what we [Japanese oilmen] think they [the oil-producing countries] need," (2) "what they think they need," and (3) "what they really need."[14] The obvious implication is that all three may differ from one another. Japanese businessmen have to be able to offer what the governments of the host countries believe they need in order to win the goodwill of the countries where they have made investments. In the eyes of the host countries, if Japan fails to do so, rebuff and resentment may occur. Even if Japan and Japanese business succeed in offering what the recipient countries think they need, the final outcome can still lead to undesirable results if the consequences of economic and political developments to which the successful Japanese investments will have contributed are contrary to the long-term benefits of both sides, not to mention any side effect involving third parties.

Some Japanese seem to be of the opinion that the paramount concerns of the host countries where Japan hopes to find oil are the extent of participation and the manner in which profit will be shared, the investment obligations of Japanese oil companies, and the magnitude of the "signature bonus."[15] According to this view, acquisition of new technology from Japan or, for that matter, from any other foreign investor may be quite low on a potential host country's priority list. This particular view may well have been based upon individual experiences during competitive bidding for concessions or exploration rights in some countries. According to Japanese oilmen who have operated in such countries as Saudi Arabia, Kuwait, Nigeria, and Indonesia, the governments of host countries have always been interested, albeit to varying degrees, in the employment of nationals rather than persons brought in from Japan, in participation in management by local personnel, in reinvestment of earnings by the companies in the same host countries, in concrete expressions of concern in the host countries' general development through company-sponsored educational, training, and social programs, and in practical expressions of corporate social responsibility. Depending upon the individual nations' particular stages of development and the background and training of their economic planners and lower-level officials, transfer of technology through conscious training efforts and through importation and employment of the latest machines and their embodied technology is also regarded in some countries—definitely in Southeast Asia—as a most important desideratum. Since Japanese technology in oil field operations lags somewhat behind Western technology, this point may appear less important to some Japanese, in the short run at any rate. It is definitely not so in investment in general.

In some of the oil-producing countries, Japanese firms have been criticized on various counts, although such censure often is based on impressions of the behavior of trading companies rather than on that of investors in the oil industry. Japanese businessmen, for instance, are accused of foisting obsolete and inefficient equipment on nations where they have capital outlay, so that what purports to be an act of investment may appear to nationals of host countries to be export of obsolete equipment. These firms are said to be extremely jealous in safeguarding sophisticated technology, a point related to another criticism: that nationals of host countries do not enjoy vertical mobility in Japanese enterprises and that they have little opportunity to learn technical and managerial skills and to be given responsible positions. Such criticism often is voiced through comparison with practices of other foreign firms, especially American companies which, in the eyes of some of the host countries, tend to do better in this regard.

Japanese businessmen are said to be aloof, keeping mostly to themselves—an attitude sometimes taken for arrogance and disinterest in the countries where they make their profit. Living in Japanese-owned hotels, frequenting Japanese-owned shops, playing golf in their own recreation facilities—such behavior is sometimes denounced as recycling of Japanese money in the sense that very little of these expenditures in foreign countries reach the local economy. Ironically, the very fact that Japanese enterprises are in many industrial fields in some of the nations where their investments abound is regarded as a potential threat to national control of the economy rather than a reflection of the broad-gauged interest in that economy held by Japanese investors.

Where participation of local personnel in Japanese firms is in question, the small number is sometimes treated as evidence of a Japanese predilection to deal through middlemen who in Southeast Asia are often ethnic Chinese. This criticism is found in countries where hostility toward the Chinese elements in the local population still plays an important part in local politics and economic policy.

Still other aspects of business management in overseas Japanese establishments, which appear to officials and businessmen in the host countries to be characteristically Japanese, include inadequate delegation of authority by the head offices of Japanese firms to their local branches, the low rank held in their home offices by Japanese personnel sent overseas, a highly structured management, and lack of an understanding relationship between management and labor.

In general, stated bluntly, there is often a deep concern that the Japanese investors may turn out to be exploiters who will extract and remove the host nations' resources, leaving little technical and material benefits when they leave and providing only a minimal amount of local employment and value added while they are in the country. Of course, many persons in these countries know that this is an unduly pessimistic picture, that it may well be an unfair picture, and that in any case some of the undesirable effects certainly can be avoided. What is equally important is that many Japanese have become increasingly concerned about these criticisms. This is true both in official circles in Japan and among businessmen in the field. Some of the oilmen, for instance, are fully aware of the characteristics of the organization and practice of Japanese management and the latter's lagging adaptation to local conditions. The very large scale of Japanese foreign investments in general and, more recently, in the oil industry, as well as the newness of these activities, have exceeded the rate at which adequately trained personnel can be supplied and recruited for work overseas. Inexperience, aggravated by what may be a characteristically Japanese single-mindedness and made worse by competitiveness in aggressive export promotion—which has become almost a habit—is probably at the root of the many situations and practices that others find irritating and unconscionable.

As newcomers to the field of overseas oil exploration, Japanese firms suffer from the disadvantage that they are expected to live up to standards already set by others in such areas as educational programs, participation in community affairs, and investment in other industries. In Saudi Arabia, for instance, the Japanese Arabian Oil Company may be compared with Aramco. Of course, the existence of certain accepted standards also makes it easier for the late arrival to emulate them. The helpfulness of some Western oil companies in showing the way is readily acknowledged by Japanese oilmen. The former are not motivated by altruism alone; they have a self-interest in not stirring up xenophobic outcries in general.

Japanese versus Western oil firms. Another point deserving attention is the relationship between Japanese oil firms engaged in overseas activities and the Japanese government. Even companies without JPDC financing seek MITI approval before any significant action is taken in other nations and prior to dealing with foreign government agencies and firms. However, are Japanese government agencies themselves geared to provide clear instructions and efficient execution? Is the usual administrative guidance predicated on wise counsel? Some Japanese oilmen seem to think that, like themselves, government agencies suffer from being relatively new at a game which is expanding so rapidly. There still is a shortage of linguistic talents and persons with the necessary cultural and technical background to deal with the oil-producing countries where Japan seeks to enlarge its investment. Others, keenly aware of the real or imagined shortcomings of their own government agencies and representatives, may even entertain an exaggerated notion of the degree of coordination between the governments and business firms of other countries. For instance, they regard major U.S. oil firms as willing and able instruments of the U.S. government. They lament their own lack of adequate support from the Japanese government.

It is but a short step from this point for them to argue further that Japan's lack of a navy to show the flag presents a serious disadvantage that must be redressed. However, the method Japanese businessmen have often used to offset their perceived disadvantage has been to try to establish intimate, friendly relations with high officials in the host countries—which may well explain the complaint of one Southeast Asian government official to this writer that every visiting Japanese businessman wishes to call on him, as if there were a company directive that this be done. Furthermore, oilmen who have to deal with local officials in isolated areas have found it useful to cultivate officials at the national level of the host government for the sake of leverage and convenience.

Potential complications and conflicts. The preceding discussion leads to several points. First, through no fault that can be specifically attributed to them—perhaps even in spite of their efforts—Japanese firms looking abroad for oil are regarded by nationals in other countries in the image already created by Japanese trading companies and other Japanese businessmen. While Japan's record in World War II is no longer publicly stressed in many countries, popular memory in Southeast Asia dies hard. Even knowledgable officials who are aware of Japan's changed military status and foreign policy do not seem entirely comfortable in freely admitting Japanese enterprise and capital. For the Japanese and Southeast Asians, a long period of learning about one another still lies ahead. It follows, therefore, that Japan is by no means assured of unquestioned welcome and success in its effort to expand the autonomous supply of oil it needs through Japanese-operated, if not Japanese-owned, enterprises. This is true especially in such countries as Indonesia, Malaysia, and Thailand, where Japanese oil exploration should expand in the interest of geographical diversification to the east of Malacca. Anti-Japanese demonstrations during Tanaka's 1974 visits to two of these three countries were a well-publicized reminder of a deep-seated problem.

Given the nature of the potential and actual resistance to Japanese capital and enterprise, it would seem that a most important condition of success is time. Time is needed to change attitudes and build trust. Japanese promises of more money for foreign aid to the developing countries may be another necessary ingredient of success, but it is definitely not

TABLE 23

INTENDED MARKETS OF JAPANESE INVESTMENTS
IN MANUFACTURING IN FOREIGN COUNTRIES

(in percentage of cases reported)

Number of Cases Reported		Safeguarding of Existing Local Markets in Host Countries	Sales in Expanding Markets in Host Countries	Participation in New Markets in Host Countries	Export to Third Countries	Export to Japan	Other	Total
Before 1965	297	24	39	18	11	4	4	100
1966–1969	450	20	31	13	18	13	5	100
1970–1971	381	19	28	19	16	17	1	100

SOURCE: *The Current Conditions and Prospects of Our Country's Overseas Investment.* The Fourth Overseas Investment Survey Report. Published by the Japan Export-Import Bank, 1974. For reference, see also *Japanese Private Investments Abroad* (summary of the third survey report), in English, October 1972.

sufficient alone. As a matter of fact, an overaggressive approach may produce exactly the opposite result. The more Japan feels compelled to speed up its diversification and autonomous supply program for political reasons, the greater is the chance of failure.

Time, however, may also work against improving relations, between Japan and the oil-producing recipients of Japanese investment. In four successive surveys of Japanese private foriegn investment conducted by the Japan Export-Import Bank since 1965, the respondents were asked to state their main reasons for investing abroad.[16] The dominant considerations in the earlier surveys were very clearly protection of existing export markets and participation in anticipated market expansion, both in the host countries and elsewhere, under conditions of increasing protectionism which the respondents saw developing. In the survey of 1971–72, the trend had shifted somewhat, probably reflecting both rising labor costs and an expanding domestic demand in Japan (see Tables 23 and 24). More Japanese investors now stated as their main reason for overseas investment the possibility of production for export to Japan using the much cheaper labor of the host countries. If, as time passes, Japan's need for foreign exchange earnings should increase because of the high oil-import bill, the export emphasis—which had never really decreased very much—may again reassert itself. This could become another irritant in bilateral relations between Japan and the host countries of Japanese capital.

While one advantage of investing in the development of natural resources abroad, such as oil, is to secure supply for Japan, another benefit may

TABLE 24

ADVANTAGES OF OVERSEAS INVESTMENT AS PERCEIVED
BY JAPANESE MANUFACTURING FIRMS

Number of Cases Reported		Host Countries' Policy of Protection for Industry	Abundant Supply of Labor	Abundant Supply of Raw Materials and Energy Resources	Abundant Supply of Land and Water	Proximity to Market	Trade Advantage between Host Country and Third Countries	Other	Total
1965	367	32	25	8	4	19	9	3	100
1966–1969	527	25	36	5	5	16	10	3	100
1970–1971	461	21	36	7	4	17	11	4	100

SOURCE: Same as Table 23.

lie in the possibility of securing more advantageous prices. If a Japanese investor is both the developer of the resource and its consumer downstream, a lower input price would tend to boost profit downstream and reduce revenue at the foreign end. If the foreign host country government or its nationals have a direct interest—for example, because tax revenue or profit sharing or payments to employees are believed to be affected—in the division of benefits between the foreign input-producing side and the Japanese downstream interest, another irritant will be present. As a matter of fact, according to General Ibnu Soetowo, director of Pertamina until 1976, it was this very consideration that had led Indonesia to favor "production-sharing contracts" over the older "concession formula."[17] This issue is by no means peculiar to oil and other natural resources; it applies no less to manufacturing. To the host country where a firm with Japanese capital imports intermediate products from a business in Japan controlled by the same interests or, alternatively, where the former sells its products to the latter for further processing or marketing, pricing and the division of potential benefits without the intervention of open market competition is a constant source of recrimination and complaint. This issue may loom larger if the trend of Japanese private foreign investment for import to Japan should continue.

Two additional complications may arise in the long run. In the first place, relations between Japanese and U.S. oil firms and between Japanese and U.S. businesses in general may become more strained as a result of increased competition between the two parties. What is true of Japanese-U.S. business relations would apply equally to Japanese relations with businesses owned by other Western countries. Mention has already been made of the generally cooperative relationship between U.S. and Japanese oil firms in a number of joint ventures in oil-producing countries. The Japanese still need to learn U.S. technology at the present stage of Japan's overseas expansion in oil production. Both Japanese and U.S. firms have found it desirable to share financial commitments and risks. Both have felt a kinship in the face of the increasing demands of the producing countries. However, even now some Japanese oilmen resent their own "poor cousin" position as nonoperators that provide only financial support to American firms operating in the field. As time passes, this cooperative relationship too may not endure.

Whether serious strain would develop may depend upon how both Japan and the United States solve their international payments problems vis-à-vis increased petroleum costs. Looking at the problem from the Japanese side, if payments for a continuously larger oil bill cannot be met fully by (1) rising prices and volume of Japanese exports to the oil-producing countries in the current period, (2) Japanese investments in goods and services in advance of future oil imports and future oil bills, and (3) credit extensions—either short- or long-term—by the oil-producing countries to Japan, then the balance would have to be made up by greater current export to other oil-importing countries, transfers of real assets in Japan, and/or similar transfers of Japanese-owned assets in other countries to the oil-producing countries. Since Japan would probably be unwilling to permit unlimited transfer of both ownership and control of domestic industrial assets and real estate, the last method of payment could become quite important if portfolio investment in minority holdings do not satisfy the investors with oil money. This means that Japan would have to acquire assets in other countries for transfer to the oil-producing countries, and these assets must be in forms acceptable to the latter. The most readily acceptable would be convertible currencies which can be earned through the accumulation of export surpluses in trading with third countries.

All this points to a renewed Japanese export drive both in the industrial West and in the less-developed countries. This export drive would not be able to limit itself to additional Japanese exports derived from an expansion of the foreign trade of other countries. Part of its effect may well be the replacement of exports from other industrialized economies in the less-developed areas of the world—an effect which could be concentrated in some of the less-developed oil-producing countries, especially because the Japanese effort will be focused on them. Thus we could envisage an exacerbation of relations between Japan and such countries as the United States and members of the EEC. In Southeast Asia, where a convergence of U.S. and Japanese economic interests is needed to support a convergence of general foreign policy which the two countries would seem to desire, such a development would be very harmful to both sides. Thus we come to the real basic issue in the Japanese policy to become more self-reliant in its oil supply. Is it at all possible in the long run to achieve such independence, including financial independence with respect to oil payment, or will the attempt to reach this goal merely lead to new frustrations and undermine other aspects of Japan's national security?

The recycling issue could be made easier if the oil-exporting countries having capital to invest would be content with portfolio investment. There is a possibility that some very small OAPEC countries such as Kuwait and Abu Dhabi, which could never aspire to becoming big powers, may share this attitude or could be persuaded to do so. Kuwait, which acquired five million shares of New Japan Steel (one-thousandth of the outstanding stock) in early 1975,[18] is said to be following four general principles in placing foreign investments: (1) preference for large enterprises and observance of local laws; (2) not insisting on management participation in case of disagreement with interested parties in the host country; (3) acquisition of stocks for long-term investment rather than speculation; and (4) cooperation with host countries in promoting social welfare. A priori, Japan should have no qualms in accepting foreign capital on this basis. There is a special advantage in dealing with small investors interested primarily in future income security rather than political prestige. It is not clear whether Saudi Arabia can be included in this category of smaller countries. Obviously, one would not count Iran among them. Unfortunately, however, whether the small oil-producing countries are politically stable and viable as independent entities is an open question.

Will development bring stability? Finally, we must examine another potential pitfall that may actually be the net result of initial success in the Japanese effort to expand oil and other investments abroad. Let us assume for the sake of argument that the oil-producing countries will continue to welcome Japanese investment and that economic growth measured in GNP and industrial output manages to take place. Let us, however, also raise the question whether such economic growth will necessarily be stabilizing. The answer to this question will probably depend upon the speed at which the population's expectations will rise and will change as they rise, and upon the ability of the local government to satisfy these changing—probably growing—expectations in terms of employment levels, educational opportunities, and social and political welfare and aspirations. It will also depend upon the nature and wisdom of the political elite that will emerge with economic growth, aided by foreign technical assistance and foreign capital. In particular, as the oil-producing countries develop economically, will new elite groups such as the military emerge, and will they be willing to accept the leadership of the existing political parties and groups now in power? Will the latter be wise and able enough to transfer power by peaceful means where transfers cannot be avoided? Will not other dissident and militant political parties, possibly egged on by interested foreign powers, contest for leadership? In an atmosphere of discontent and instability, foreign investment interests can easily become both the target and the victim. In that event security of oil supply may turn out to be illusory after all. What all this means is that independence in oil supply, in the sense of national control of sources overseas in Japanese hands, perhaps can never be fully attained or even attained to a high degree.

Oil Transportation and the Sea-lanes

We turn last to the obvious need to bring oil from the producing countries to the shores of Japan. Strictly speaking, national control of oil supply should apply to delivery to Japan no less than to production at the source. The number of Japanese tankers increased from 151 in 1970 to 172 in 1973, including 60 vessels built during the intervening period.[19] Since the new vessels are larger, the aggregate size of the Japanese tanker fleet rose from 13,143 thousand dwt in 1970 to 21,870 dwt in 1973. The average tanker size of the entire fleet in 1973 was 127,000 dwt; that of the new tankers, 202,000 dwt. However, in spite of the increase in numbers and tonnage, the lifting ratio of Japanese tankers actually declined from 1970 to 1972. Of the crude oil, fuel oil, and naptha imported in 1970, 60.2 percent were brought in by Japanese tankers and 39.8 percent by foreign vessels. The corresponding ratio in 1972 was 56 percent Japanese to 44 percent foreign. On the eve of the oil crisis, a sizable government budget was provided in the country's twenty-ninth shipping program to construct ocean-going vessels, including 2.9 million dwt of oil tankers.

From the standpoint of security, the advantage of the increase in tanker tonnage would be compounded by a reduction of days at sea. This would accrue if Indonesia and Malaysia took over a larger portion of the Middle Eastern supply. On the other hand, shifting to Indonesian and Malaysian sources of oil would substitute the influence of Indonesia and Malaysia over the Malacca Straits. An alternative to the Malacca Straits is Lombok; here Indonesia alone would exercise the controlling influence. The cost of such a substitution would be a small increase in the number of days at sea. Still another potential alternative is a canal or pipeline across the Kra Isthmus. The project was revived a few years ago, and a Japanese trading company is said to have shown interest, as has Saudi Arabia.[20] However, after the communist takeover of all Indochina and the reassessment by Thailand of its own external posture, this potential northern route across the South China Sea has lost a great deal of its attractiveness. The security aspect on this route is complicated by the unpredictable behavior of the new communist states. To the north Japan could shorten its oil route even more by buying oil from the PRC or the Soviet Union. The political leverage such a shift would give to the communist powers has already been discussed. Finally, there is the position of Taiwan. If it were in the hands of a government hostile to Japan, it too could seriously constrain martime traffic from the south. Even if Japan could find a great deal more oil immediately offshore, it would still not be able to bypass certain inherent political and military issues. The Senkaku Islands (or Tiao-yü-t'ai) and areas between Korea and the PRC are all territories in actual or potential dispute.[21]

The crux of the issue is whether, unaided by others, Japan can safeguard the sea-lanes, regardless of their length, and protect offshore drilling operations. In the absence of agreement among the major powers and the relevant regional states, one must seriously question the ability of Japan to provide protection on its own. The respective capabilities and future roles of the U.S. Seventh Fleet and the Soviet Pacific Fleet in the Indian and Pacific Oceans are among the principal uncertainties from the Japanese point of view.

5

CONCLUSION

In the face of the high price of oil demanded by OPEC and the recent experience of a threatened cutback of supply by the Arab countries, Japan has sought to redouble its efforts to gain a greater degree of national control over sources of supply, the intermediaries linking the centers of production with the Japanese market, means of transportation, and Japan's ability to pay for the unavoidable import. In addition, Japanese planners have indicated that they anticipate during the 1980s a smaller oil demand, a lower growth rate, a gradual shift to other sources of energy, and some restructuring of domestic demand.[1] All these developments, however, cannot really obviate the program of diversification, overseas investment, and bilateral negotiations with the Soviet Union, the PRC, and the non-communist oil-exporting states we have described in this study. Nor are Japanese estimates of lower growth rates and successful curtailment of energy consumption wholly dependable. Under the circumstances, what conclusion should we draw about Japan's external policy to secure the nation's oil supply?

Perhaps we should begin by stating the conditions under which one or more aspects of Japan's oil problem might disappear. First of all, the oil cartel might break up. This could be induced by declining oil prices and a further reduction of overall demand, aided by the need of individual producing countries to increase their own oil income because of unusually large actual or committed expenditures. Improvidence in spending, evidence of which began to appear in 1975, could also increase direct imports, at least for a while, from such oil-importing countries as Japan. The long-standing Arab-Israeli dispute which galvanized the Arab states into collective action could be replaced by other preoccupations among some of the Arab states themselves. Internal problems may also worsen. The expansion of national armaments could well increase the intensity of any conflict while making it more likely. Too rapid an economic expansion and rise of expectations without the necessary institutions and leadership to satisfy these demands might engender greater internal instability.

Second, the Soviet Union may continue to overplay its hand in negotiating with Japan. Peking may continue to be embroiled in its own internal political and ideological struggle so that, together with its fear of the Soviet Union, it will remain politically and militarily passive. Even its much higher oil income may continue to be wiped out by its simultaneously more costly purchases from Japan.

Third, the Soviet and U.S. fleets may continue to be balanced in favor of the latter.

Given these conditions, Japan will not be faced with a serious oil problem in the short run, and it will have time for long-term adjustments.

Unfortunately for Japan, even if these fortuitous developments were to take place ,there is no guarantee that difficulties might not arise involving other raw materials that the country must import entirely.[2] More likely, however, is the continuation of at least some of the international political and economic conditions that gave rise initially to the oil crisis.

Given the continuation of the underlying conditions of the oil crisis and the fact that success in avoiding its worst impact in 1974–75 should really be attributed primarily to the decline in demand for oil largely due to the worldwide recession, Japan will have to continue to pursue its program to increase national control of oil supply. We have seen in the preceding discussion that such an effort to diversify sources of supply and to increase direct national control over production, importation, and transportation really amounts to an exchange of one kind of risk for another and of one group of foreign countries holding leverage over Japan for another. Full independence from foreign influence is an objective that cannot be attained. To pursue such an objective is to try to capture an illusion. Realistically, therefore, there will have to be a trade-off of risks from Japan's point of view, and the end result should aim at the minimization of the impact of any untoward development.

In this connection it behooves Japanese planners to bear in mind the following considerations. First, a distinction should be made between those oil-producing countries that have no fundamental objection to international economic interdependence and those that aim at economic self-sufficiency or, at the very least, insulation of their own national economic plans from uncontrollable world events. For instance, diversion of oil import from the Middle East to Indonesia or Malaysia could lead to greater interdependence between Japan and these two countries rather than one-sided Japanese dependence. On the other hand, increasing oil import from the Soviet Union and/or the PRC would not necessarily be accompanied by Soviet or Chinese economic interdependence with Japan. The nature of the dependence is more likely to be one-sided.

In the second place, one must bear in mind that an overzealous effort to achieve independence in oil or even interdependence with the Southeast Asian countries and others could prove to be counterproductive. Japanese economic nationalism would only serve to fan the flames of economic nationalism on the part of the other nations. It is not sufficient to increase Japan's foreign aid grants to less-developed countries in order to increase their friendliness or reduce their hostility toward Japan. The behavior of Japanese businessmen in general will influence local perceptions of the benefit of economic interdependence. Both Japan and the developing countries should realize that economic development itself may be destabilizing in a broader sense and therefore may work against international cooperation if expectations generated cannot be fulfilled, and if foreigners are handy to serve as targets of nationalism and xenophobia.

What if some of the risks should materialize in spite of diversification? What if there is to be another threat of oil cutback or even an actual cutback by some of the suppliers? What if the price of oil should rise again? What if demand for oil should again increase rapidly even without any further price increase, once worldwide economic growth resumes? Will other oil-importing countries such as the United States be willing to share their oil or even become oil-exporting countries to Japan? Clearly the extent of international cooperation on energy needs to be expanded beyond the stage of sharing during an emergency, the building of emergency stocks, and long-term research and development to produce cheaper substitutes. The Washington energy conference of 1974 and subsequent IEA activities have provided a beginning, but little has been accomplished in developing real invulnerability to OPEC since the 1973 oil crisis.[3]

The performance of Japan's foreign policy during the oil crisis has raised certain questions in the minds of non-Japanese observers. However, in order to formulate an intelligent international economic policy for the benefit of *all* nations rather than a particular country alone, other countries must have equally intelligent policies. This study on Japan's oil policy has strengthened our underlying thesis about the inherent interdependence of economic and security policies in U.S.-Japanese relations. Both countries claim that each is the other's principal partner in the Pacific. This claim remains yet to be confirmed by future developments in the case of oil policy—cooperation in IEA, whose efforts are limited, notwithstanding.

Thus far, U.S. policy under Secretary of State Henry Kissinger has been to seek first a consensus among the oil-importing countries—and later agreement with the oil-exporting countries on such matters as price and recycling of oil dollars. Japan's response to this policy has been to straddle both the U.S. approach and what is essentially a national solution.[4] Agreement among the importing countries on a long-term solution has been exceedingly slow in coming[5] and agreement between exporting and importing countries will probably be equally slow. While Japan's effort to seek a national solution may be pushed forward with characteristic zeal and other complications can also arise, the task of coordinating the future policies of the two countries will not become easier as time passes.

There is an off chance that Japan's national solution may be successful. Other developed countries, including the United States, may be less alert and able in competing with Japan in promoting exports and foreign investment to safeguard oil and other supplies. Oil-producing countries may become more receptive to Japanese overtures, perhaps precisely because Japan has no military means to impose its will on them, than they would be to U.S. and European overtures. Such a scenario does not, however, appear to be very plausible in the long run.

APPENDIX

The following two tables are presented to show the progressive downward adjustment of supply targets for oil and all energy products after the 1973–74 oil crisis. These changes reflect downward adjustments of demand and economic growth, reinforced by energy conservation. However, apart from the smaller percentage shares assigned in the later estimates to nuclear power, which may well be correct, the rest is probably more indicative of hopes and intentions than of realistic projections. If the proportion of nuclear power drops further as a result of environmentalists' objections and site problems, the demand for oil would rise accordingly, short of a far more rapid rise of coal import and supply of other nonfossil fuels. For more details see Chapter 1, last section.

TABLE A

PETROLEUM SUPPLY PLANS OF JAPAN

(in thousand Kl) (Petroleum gas – 1000 tons)

	Pre-crisis (1973) 5-year Plan for FY 1977 (1)	Post-crisis (1974) 5-year Plan for FY 1977 (2)	Projection in 5-year program (1975) for FY 1977 (3)	(2) in percentage of (1)	Change in percentage	(3) in percentage of (2)	Change in percentage
CRUDE:							
Indigenous production	1,000	1,000	1,000	100.0	–	–	–
Imported for refining use	332,326	307,445	279,413	92.5	−7.5	90.9	−9.1
Imported for nonrefining use	1,434	1,749	3,651	122.0	+22.0	208.7	+108.7
Total	335,760	310,194	284,064	92.5	−7.5	91.6	−8.4
PETROLEUM PRODUCTS:							
Produced							
Gasoline	36,257	31,570	30,676	87.1	−12.9	97.2	−2.8
Naphtha	40,078	39,015	32,467	97.3	−2.7	83.2	−16.8
Jet fuel oil	6,050	4,787	4,395	79.1	−20.9	91.8	−8.2
Kerosene	30,595	27,404	24,841	89.6	−10.4	90.6	−9.4
Gas oil	22,472	20,704	19,177	92.1	−7.9	92.6	−7.4
Heavy fuel oil	169,888	158,860	141,016	93.5	−6.5	88.8	−11.2
Subtotal	305,340	282,340	252,572	92.5	−7.5	89.5	−10.5
Petroleum gas	5,905	5,423	4,813	91.8	−8.2	88.8	−11.2
Imported (excluding bonded)							
Naphtha	6,000	4,000	4,000	66.7	−33.3	–	–
Heavy fuel oil	45,715	37,724	33,979	82.5	−17.5	90.1	−9.9
Subtotal	51,715	41,724	37,979	80.7	−19.3	91.0	−9.0
Petroleum gas	8,564	8,845	8,475	103.3	+3.3	95.8	−4.2
Bonded							
Jet fuel oil	1,034	770	660	74.5	−25.5	85.7	−14.3
Heavy fuel oil	1,267	3,110	3,065	245.5	+145.5	98.6	−1.4
Subtotal	2,301	3,880	3,725	168.6	+68.6	96.0	−4.0

SOURCES: (1) *The Petroleum Industry in Japan*, 1973, the Japanese National Committee of the World Petroleum Congress, p. 16.
(2) *Comprehensive Energy Statistics*, 1974, p. 221, revised figures.
(3) MITI, reproduced in Institute of Energy Economics, *Energy in Japan* (June 1975), p. 7.

TABLE B

LONG-TERM EXPECTATIONS FOR THE 1985 FISCAL YEAR

	As of July 1970	As of June 1974	As of June 1975
Annual GNP growth rate (in percent)	*1976–1985* 8.5–9.5	*1973–1985* 6–8	*1973–1985* 5.5
			1975–1985 6
Total energy supply in oil equivalent (million kiloliters), 1985	933.3–1,028.9	727–915	760–830
Percent distribution			
1. Hydropower	2.5–2.3	3.4–3.2	3.7
2. Geothermal power	.....	0.3–1.1	0.5
3. Nuclear power	9.9–9.1	10.3–11.4	9.6
4. Coal			
Domestic	2.2–2.0	1.9–1.5	1.9
Import	14.5–14.7	11.0–9.4	11.2
Total	16.7–16.8[a]	12.9–10.9	13.1
Subtotal of 1–4	29.1–28.2	26.9–26.6	27.1
5. Domestic oil and natural gas[b]	2.3–2.0	1.9–3.1	1.8
6. Imported petroleum (crude oil and products)	67.2–68.5	64.6–61.4	63.3
7. Imported liquid natural gas (LNG)	1.4–1.3	6.6–8.9	7.9
Total	100.0 100.0	100.0 100.0	100.0
Import			
Crude oil (million kl)	649.6–723.4		
Petroleum products (million kl)	56.2–63.5	500–600	485
LPG (million tons)	6.7–8.2		

SOURCE: The Energy Advisory Council (MITI) July 1974. The 1975 data are provided by the Institute of Energy Economics. The difference assumes a 9.4% reduction of energy demand through conservation.

[a]Components do not add up to total due to rounding.

[b]Derived as a remainder.

NOTES

1: Background

1. Original text quoted by Stockholm International Peace Research Institute (SIPRI), *Oil and Security* (New York: Humanities Press, 1974), appendix 6, p. 118. Iraq did not cut back its production. In the case of other countries, observance of the embargo was far from being watertight. See Chapter 2, last section.

2. *Daily Yomiuri* (Tokyo), October 17, 1973.

3. Ibid., October 25, 1973.

4. Ibid., October 26, 1973.

5. Ibid. For example, such a notice was issued by the Maruzen Oil Company, a major refinery supplying oil to Tokyo, Kansai, and Chubu power plants. However, Maruzen indicated that the supply cutback would take place initially in low-sulphur oil from Indonesia which was supplied by the Union Oil Company, a firm cooperating with Maruzen.

6. SIPRI, *Oil and Security*, p. 119. The oil no longer exported to the United States and the Netherlands under the total embargo against the two countries was to be counted as a part of the 25 percent cutback. This would make the reduction for the other consuming countries actually smaller than the 25 percent figure might suggest.

   The decision on how much to reduce production and how to share the total cutback among the producers is an economic and political problem facing the oil cartel. Interestingly, the 25 percent figure was employed by Sheikh Ahmed Zaki Yamani, the Saudi Arabian petroleum minister, when he reportedly stated in April 1973, six months before the October war, that Saudi Arabia had the capability of setting a much higher world price for oil by simply reducing its current production by 25 percent. *Washington Post Service*, April 18, 1973.

7. SIPRI, *Oil and Security*, p. 120.

8. Ibid., p. 121.

9. Japan Petroleum Consultants (Tokyo), *Japan Petroleum Weekly*, vol. 9, no. 4 (January 28, 1974):2.

10. Ibid.

11. For various details of the controls introduced during this period, see ibid.,

vol. 9, no. 2 (January 14, 1974); ibid., no. 8 (February 25, 1974).

12. For the full English text of the 1975 recommendations of the Energy Advisory Council, see ibid. (September 1975) and subsequent issues.

2: Vulnerability: Theory and Fact

1. For a discussion along these lines, see Hollis Chenery, "Restructuring the World Economy," *Foreign Affairs* (January 1975).

2. *Comprehensive Energy Statistics* (Tokyo: Tsûshô Sangyô Kenkyûsha, 1974), pp. 190–91.

3. A part of the transportation use of gasoline and other personal consumption (mostly for heating) is assumed to represent direct personal consumption.

4. Calculated on the basis that oil and petroleum products constitute 75 percent of total primary energy supply.

5. Theoretically they could also do the opposite and especially favor Japan. This possibility, however, is never mentioned in Japanese discussions of the problem.

6. See *Monthly Statistics of Japan* (October 1975), p. 52; also *Tsusho Hakusho* [MITI White Paper] (1974), pp. 164–65.

7. These statistics were compiled by the Energy Advisory Council in July 1974.

8. *Comprehensive Energy Statistics* (1974), p. 189.

9. *Japan Petroleum Weekly*, vol. 9, no. 11 (March 18, 1974):2.

10. Ibid., p. 1.

11. Japan Foreign Trade Council report in March 1975. The 1973 figure is taken from data given by the Energy Advisory Council in July 1974. See also *Monthly Statistics of Japan* (March 1975).

12. An extract of the report of the Japanese Economic Planning Agency is given in *News from MITI* (Tokyo: MITI Information Office, April 21, 1975). Oil import in 1974 was at 278 million kiloliters or $18.9 billion. The merchandise trade figures used here are the unadjusted original data of the Bureau of Customs given in *Monthly Statistics of Japan* and differ from the seasonally adjusted figures in Table 10.

13. See *Wall Street Journal*, September 17, 1974; *San Francisco Examiner*, August 14, 1975; *Wall Street Journal*, September 29, 30, 1975; October 7, 17, 20, 1975; November 3, 9, 1975; *San Francisco Chronicle*, February 16, 1976.

3: Interaction with Moscow and Peking

1. Kazuo Ogawa, *Shiberia Kaihatsu to Nihon* [The Development of Siberia and Japan] (Tokyo, October 1974), part I, chapter 5, 6; and part 3, chapter 3.

2. *Comprehensive Energy Statistics* (Tokyo, 1974), p. 199.

3. See, for instance, Vaclav Smil, "Energy in the PRC," *Current Scene*, vol. 13, no. 2 (February 1975); R. M. Field, "Chinese Industrial Development; 1949–70," *People's Republic of China: an Economic Assessment* (PRC:EA, 1972), table B–1, p. 83. For an historical account of Chinese oil, see H. C. Ling, *The Petroleum Industry of the People's Republic of China* (Stanford, California: Hoover Institution Press, 1975).

4. Ogawa, *Shiberia Kaihatsu to Nihon*, p. 156; original sources from Soviet foreign trade statistics and MITI.

5. Estimated by U.S. analysts at $1.108 billion out of a total turnover of $5.83 billion.

6. Calculated at $52.1 billion for Japan's global foreign trade (export plus import) in 1972 and total Sino-Japanese trade at 338.8 billion yen at an average exchange rate of 308 yen per dollar. *Monthly Statistics of Japan*, no. 166 (April 1975), pp. 50, 53.

7. For a discussion of U.S. policy in this respect, see the author's study on *U.S. Policy and Strategic Interests in the Western Pacific* (New York: Crane, Russak and Co., 1975).

8. Jetro, *China Economic Research Monthly* (March 1974), p. 55. Comparable figures are given by Ogawa, *Shiberia Kaihatsu to Nihon*, table III–6, p. 163.

9. See Ogawa, *Shiberia Kaihatsu to Nihon*. In addition to the above-mentioned projects, a wood pulp and chip project was discussed during the third meeting in February 1968. Bilateral discussions have also covered such topics as copper, iron ore, and the construction of an oil refinery at Nakhodka.

10. Reported in *Yomiuri*, February 15, 1973.

11. Ibid., March 7, 1973.

12. *Nihon Keizai* (Tokyo), March 13, 1973.

13. *Mainichi* (Tokyo), March 14, 1973.

14. *Yomiuri*, March 16, 1973.

15. *Nihon Keizai*, April 14, 1973.

16. It should be noted that Idemitsu was also an active member of the Japanese group negotiating with Peking.

17. *Nihon Kogyo* (Tokyo), June 6, 1973.

18. Ibid., Aug. 17, 1973. For crude oil, each kiloliter equals approximately 0.875 ton; each ton equals 1.142 kiloliters. Ogawa, *Shiberia Kaihatsu to Nihon*, p. 252. Press reports often use kiloliters and tons interchangeably, because of the approximate 1 to 1 ratio.

19. The full committee did not seem to have met officially in 1973.

20. See *Sankei* (Tokyo), June 18, 1973, and *Yomiuri*, June 27, 1973.

21. *Tokyo Shimbun*, September 11, 1973, and *Asahi* (Tokyo), September 6, 1973.

22. See *Yomiuri*, September 8, 1973.

23. For a good discussion of the size of Soviet reserves and production, as well as future prospects, see Marianna Slocum, "Soviet Energy, an Internal Assessment," *Technological Review* (MIT), vol. 77, no. 1 (October–November 1974): 17–33. For popular Japanese speculation along the same lines see *Yomiuri*, March 27, 1974. For a contrary view on the trend of Soviet demand, see Ogawa, *Shiberia Kaihatsu to Nihon*.

24. Discussions on these points appeared, for example, in *Asahi*, September 6, 1973; *Sankei*, September 27, 1973; and the *Nikkan Kogyo*, September 29, 1973.

25. See *Yomiuri*, October 18, 1973.

26. See *Asahi*, March 28, 1974, and Ogawa, *Shiberia Kaihatsu to Nihon*, pp. 222–25; also *The New York Times*, April 28, 1974, Sunday Edition, Section 4, p. 5.

27. Report in *Yomiuri*, April 6, 1974.

28. Uemura and Nagano reportedly stated, "Businessmen and industry do not have to worry about the possibility of the proposed new railway being utilized for military purposes. All they have to do is to think about the economic development." Quoted in *Yomiuri*, March 29, 1974.

Cf. the view expressed by Donald C. Hellmann, "Changing American and Japanese Security Roles in Asia: Economic Implications" in *The Japanese Economy in International Perspective*, Isaiah Frank, ed. (Baltimore and London: The Johns Hopkins University Press, 1975). Hellmann writes,

> Despite the vision of intimate business-government collaboration conjured up in the image of Japan, Incorporated, the role of business in foreign affairs decisions is ambiguous and varies widely from issue to issue.... Despite these connections and the establishment of the national business organizations and committees to deal with specific foreign affairs issues ..., there is no clear mutual understanding regarding the procedures through which business opinion should be brought into the policy-making process. *Nor is there automatic agreement on the goals of the nation's* foreign policy. [Italics added.]

29. Under their arrangement Idemitsu would import one-half million tons of low sulphur crude from oil produced in Abu Dhabi under the control of British Petroleum while the latter would be compensated in return with Soviet crude from the Black Sea. This arrangement is similar to previous swap arrangements which have been made necessary by the closing of the Suez Canal. See *Nihon Keizai*, December 8, 1974.

30. For details of the several projects see *Sankei*, January 31, 1974; *Nihon Keizai*, November 5, 1973 and April 4, 1974; *Nihon Kogyo*, March 6, 1973; and *Yomiuri*, April 22, 1974.

31. Reported in *Mainichi* (evening), April 30, 1974.

32. The announcement was made by Sumita of the Japanese Export-Import Bank on January 17, 1975. See *Yomiuri*, January 18, 1975.

33. See *Nihon Keizai*, April 27, 1974. The paper also reported that the price reduction would be continued during the first ten years of production.

34. For additional details on the Sakhalin Agreement see *Yomiuri*, December 10, 1974, and January 9, 1975; also *Asahi*, January 29, 1975; and a *Kyodo* report from Moscow, December 11, 1974.

35. See *Mainichi*, April 1, 1974.

36. See *Asahi*, September 6, 1974. The same report also mentioned that the Soviet Union had proposed that it would retain only 6 of the 12 million tons of the Nakhodka refinery's output for its own use, leaving the balance for export to Japan. It is not clear from this report what role the small refineries at Khabarovsk would play in the total balance.

37. Moscow Radio in Japanese, December 12, 1974.

38. See *Mainichi*, November 17, 1974.

39. The estimate adopted by H. C. Ling in *The Petroleum Industry of the People's Republic of China*, table II, p. 15, is 5.5 million tons. A lower estimate of 4.6 million tons is given by Vaclav Smil, "Energy in the PRC." For the original sources used by Smil see "Energy Developments in the People's Republic of China," *Current Scene* (February 1975), table I, p. 3.

40. *Nihon Keizai*, April 18, 1973.

41. In early 1973 low-sulphur Minas oil was quoted at $2.96 a barrel and the Chinese wanted $3.60–$3.80. By March 1973, the price of Minas had risen to $3.73. Hence the Chinese price had become competitive. See *Nihon Keizai*, April 18, 1973.

42. *Sankei*, May 21, 1973.

43. See ibid., August 22, 1973.

44. See *Asahi* (evening), June 21, 1973.

45. Ibid., September 15, 1973. Completion of the Ta-ch'ing–Ch'ing-huang-tao pipeline was announced by Peking in January 1975.

46. This announcement was reported in *Nihon Keizai*, September 25, 1973. The port expansion at Dairen to accommodate 50,000-ton tankers was expected to be completed in June 1974.

47. Ibid., May 14, 1974; and *Asahi*, February 2, 1974.

48. The membership included: (1) trading companies—Mitsui Bussan, Mitsubishi Shoji, Sumitomo Shoji, Toko Bussan, Wako Trading, and Industrial Trading; (2) oil companies—Mitsubishi Oil, Japan Oil, Toho Oil, Tohoku Oil, Seibu Oil, Far East Oil, Sea of Japan Oil, Kyushu Oil, Asian Oil, Nansei Oil, and Fuji Kosan. *Nihon Keizai*, March 14, 1974.

49. *Sankei*, August 16, 1974.

50. For an estimate of long-term Chinese oil and energy prospects see the author's article in *Current History* (July 1975). Furthermore, according to a

report in *Sankei*, August 16, 1974, credited to Hasegawa, production at Ta-kang and Sheng-li was then at the level of 20,000–30,000 barrels per day At 365 days a year and 7 barrels per ton these figures would give a combined output of not not more than 1 to 1.5 million tons to the second and third largest fields in China after Ta-ch'ing. The total output projected for 1975 given by Hasegawa would be 30 million tons higher than in 1974 (100 million tons in 1975 versus 70 million tons in 1974). If the combined output of Sheng-li and Ta-kang in midyear 1974 was still at the low level reported in *Sankei*, it is difficult to see how increased production at these two centers, together with further expansion at Ta-ch'ing, could hope to approximate the 100-million–ton level in 1975.

51. See *Yomiuri*, October 22, 1974.

52. Reported in *Sankei*, November 4, 1974.

53. *Nihon Keizai*, December 9, 1974.

54. Acronym for Perusahaan Negara Pertambangan Minjak dan Gas Bumi Nasional [State Company, National Oil and Natural Gas Mining], Indonesia.

55. Reported in *Sankei*, December 10, 1974.

56. *Nihon Keizai*, January 14, 1975.

57. See *Sankei*, January 26, 1975. Even at FOB dollar prices Chinese oil was still 20 cents a barrel more than Indonesian crude. At the same time the shorter distance between China and Japan was offset by the smaller capacity of the tankers used. In this connection a new jetty completed at Ch'in-huang-tao was reported by *Tokyo Shimbun*, January 29, 1975. Another was under construction. Thirty-thousand-ton tankers are able to use the first one while 50- to 70-thousand-ton tankers can be accommodated at the second.

For an explanation of the role of the dollar-yuan exchange rate in the price differential between Ta-ch'ing and Minas, see *Japan Petroleum Weekly*, August 4, 1975.

58. According to the returns of the Ministry of Finance, *Japan's Exports and Imports, Country by Commodity* (December 1974), Japan imported 4.53 million kiloliters of Chinese crude oil in the 1974 calendar year. At 875 kgs. per kiloliter, this is equivalent to just under 4 million tons.

59. Reported in *Tokyo Shimbun*, February 3, 1975.

60. See *Asahi*, February 1, 1975.

61. The Associated Press reported on July 8, 1975, that PRC crude oil production in the first half of 1975 rose by 74 percent over that of the corresponding period in 1974.

62. *Jetro China Newsletter* (Tokyo), no. 7 (April 1975). The estimates are authored by Masahiko Ebashi, of Jetro's Economic Information Dept. The article appeared originally in *Movement of Chinese Economy in 1974; Focusing on China's Oil and its Related Industries* (May 1975), published by the Japan–China Economic and Trade Association.

63. The specific assumptions are as follows:

Projection 1

The PRC's energy production in 1974 by source is assumed to be as follows: coal, 67.7 percent; petroleum, 21.7 percent; natural gas, 6.8 percent; hydroelectric power, 1.2 percent; fuel wood and other, 2.5 percent. This pattern, according to Jetro, differs substantially from that given by Nickolas Ludlow, "China's Oil," *U.S.–China Business Review*, vol. 2, no. 1 (January–February 1974), who presented the structure of Chinese energy demand in 1973 as coal, 85 percent; petroleum, 12 percent; hydroelectric power and other, 3 percent.

The annual rates of growth of production of the energy sectors from 1974 on are assumed to be as follows: coal, 5 percent; petroleum, 20 percent; natural gas, 20 percent; hydroelectric power, 14 percent. (Note that the 20 percent growth rate assumed for oil is the same as that given by Kumagai.)

On the demand side, the following assumptions are made: average annual population growth, 1.9 percent; average annual growth rate of agricultural output, 2.9 percent; average annual GNP growth rate, 8.3 percent.

The energy consumption coefficient, defined as numbers of BTU per U.S. dollar of value added, is assumed to vary as follows: 9,000 BTU in 1974 to 22,000 BTU in 1985; per capita consumption for general consumption purposes, increasing from 3.2 million BTU in 1974 to 5 million in 1985; energy consumption in industry, remaining constant at the 1974 level of 140,000 BTU; consumption in transportation and other sectors, increasing from 40,000 BTU to 60,000 BTU.

China's crude oil price is assumed at the constant level estimated by IBRD with a premium added.

Projection 2

In this model the annual rate of growth of petroleum production is assumed to vary from 23 percent in 1975–78 to 20 percent in 1979–80, and 16 percent in 1981–85. In addition, use of energy for purposes of general consumption is assumed to be deliberately curtailed.

64. Sekiyu Kogyo Renmei [Federation of Petroleum Producers], *Sekiyu Kaihatsu Jipo* [Petroleum Exploration Bulletin] (Tokyo), no. 23 (September 1974), p. 97. For additional comments on other estimates see a forthcoming article by K. C. Yeh and the present author on "Oil and Strategy," paper presented at the June 1976 Conference on Mainland China, held in Taipei.

65. Vaclav Smil, "Energy in the PRC."

66. Japan's import of Chinese oil amounted to 8.8 billion yen in 1973 and 120.3 billion yen in 1974, according to Japanese customs returns.

4: Effort to Develop "Autonomous" Supply

1. Petroleum Producers Federation (PPF), *Petroleum Exploration Bulletin*, vol. 9, no. 23 (September 1974), p. 66.

2. Ibid., pp. 68–69.

3. PPF, *Data on Oil Development* (1974), p. 4.

4. PPF, *Petroleum Exploration Bulletin* (September 1974), p. 68.

5. For a description of JPDC activities see *Japan Petroleum Development Corporation Law* (Law No. 99, 1967, as revised by Law No. 38, 1973), and *Outline of Japan Petroleum Development Corp.* (January 1975), both published by the JPDC.

6. See PPF, *Petroleum Exploration Bulletin* (September 1974), pp. 54–55.

7. PPF, *Data on Oil Development* (1974), p. 13.

8. Cf. the Chase Manhattan Bank, *Capital Investments of the World Petroleum Industry* (New York, 1971).

9. These countries and other political entities are Saudi Arabia and Kuwait (the Neutral Zone), Indonesia, Malaysia, Canada, United States, Abu Dhabi, Qatar, Egypt, Zaïre, Colombia, Iran, Nigeria, Australia (including Papua–New Guinea), Gabon, Peru, Madagascar, Thailand, Iraq, Mauritania, Bangladesh, and Burma.

10. During the first half of 1974 only the following Japanese-owned companies produced any sizable amount of oil:

Company	Production (in million kiloliters)
JAPEX Indonesia	2.9
Abu Dhabi Oil	0.3
Japan Petroleum Development	13.8
Total	17.0
Arabian Oil	11.1
C Itoh	1.1
Total	12.2

The first three companies have JPDC financing; the last two are privately financed. Arabian Oil is the oldest in the field.

11. According to *Japan Times*, July 21, 1975, JPDC was contemplating the establishment of a new company with private interests in order to tap foreign oil resources more efficiently. According to this source, total overseas investment expenditure by Japanese oil companies was expected to reach 700 billion yen by 1978, but the anticipated supply to be imported from these external ventures would be no more than 22 to 25 million kiloliters of crude oil a year. MITI would like to raise the share of Japanese-controlled foreign oil from development projects abroad to 30 percent of the major effort, and the great scale of operations envisaged are intended to achieve this result. Japan is also thinking in terms of developing a new Japanese-owned "international major."

12. British Petroleum, *Statistical Review of the World Oil Industry*, (London, 1973).

13. The data on holdings are derived from various reports of PPF and from *Industrial Groupings in Japan* (1973), compiled by Dodwell Marketing Consultants, Tokyo, as well as personal interviews.

14. See PPF, *Petroleum Exploration Bulletin* (September 1974).

15. This is the monetary compensation paid by the oil company before the host country will grant it the right of exploration.

16. See *Waga Kuni Kaigai tôshi no genkyô to tenkai hôkô, Dai 4-kai kaigai tôshi ankêto chôsa hôkoku* [Current conditions and prospects of our country's overseas investment. The fourth overseas investment survey report] (Japan Export-Import Bank, 1974). For comments on Japanese foreign investments from the perspective of Southeast Asia, Kernial Singh Sandhu and Eileen P. T. Tang, eds., (The Institute of Southeast Asian Studies, Singapore University Press, June 1974).

17. Lieut. Gen. Ibnu Soetowo, "Prospects of Oil Exploration in Indonesia," a lecture before the IKIP [Scientific Education Institute], Bandung, on March 14, 1972, reprinted in the *Indonesian Review of International Affairs* (Jakarta), vol. 1, nos. 3 and 4 (July 1972–73–74): 72–79.

18. *Mainichi*, March 5, 1975.

19. The data are taken from the *Petroleum Industry in Japan* (Tokyo: Japanese National Committee of the World Petroleum Congresses, 1973), pp. 32–33. Only 15 percent of the tankers are owned by Japanese oil companies; the rest are largely owned by "designated shipping companies" that enjoy preferential treatment in low-interest government financing.

20. For a brief description of the Kra proposal see K. Y. Chow, "The Kra Canal Project, Its Relation to the Economic Development of Thailand," a paper presented at the Conference of Business Opportunities in the Pacific Basin, October 1973.

21. Japan's delay in ratifying its agreement with the Republic of Korea on the demarcation of territories for oil exploration reflects its concern about this problem. The Gulf of Thailand presents another area of potential dispute.

5: Conclusion

1. This is reflected in lowered projections of energy supply and demand for oil. See Appendix.

2. See the 1974 *Economic White Paper* and the *Trade White Paper* by MITI.

3. For a discussion of various aspects of U.S. policy see the Federal Energy Administration's report, *Project Independence* (November 1974).

4. For a similar viewpoint see Joseph A. Yaeger and Eleanor B. Steinberg, *Energy and U.S. Foreign Policy* (Washington, D.C.: The Brookings Institution, 1974), pp. 158–60.

5. In a speech before a Stanford audience, October 1975, former Secretary of the Treasury George Schultz pointed to the long time lag between economic developments and policy decisions by a big government to meet any challenge as one of the most serious problems facing us today. The development of an oil policy for the U.S. after 1973, or failure to do so promptly, is a classical example.

INDEX